Elite • 39

The Ancient Assyrians

Mark Healy · Illustrated by Angus McBride

Consultant Editor Martin Windrow

First published in Great Britain in 1991 by Osprey Publishing,
Midland House, West Way, Botley, Oxford OX2 0PH, UK
443 Park Avenue South, New York, NY 10016, USA
E-mail: info@ospreypublishing.com

Transferred to digital print on demand 2008

First published 1991
11th impression 2007

Printed and bound by Cpod, Trowbridge

A CIP catalogue record for this book is available from the British Library

ISBN: 978 1 85532 163 2

Editorial by Martin Windrow

FOR A CATALOGUE OF ALL BOOKS PUBLISHED BY OSPREY MILITARY
AND AVIATION PLEASE CONTACT:

NORTH AMERICA
Osprey Direct, c/o Random House Distribution Center, 400 Hahn Road, Westminster, MD 21157
E-mail: info@ospreydirect.com

ALL OTHER REGIONS
Osprey Direct UK, P.O. Box 140 Wellingborough, Northants, NN8 2FA, UK
E-mail: info@ospreydirect.co.uk

www.ospreypublishing.com

Artist's note
Readers may care to note that the original paintings from which the colour plates in this book were
prepared are available for private sale. All reproduction copyright whatsoever is retained by the
Publishers. All enquiries should be addressed to

Scorpio Gallery, PO Box 475, Hailsham, East Sussex, BN27 2SL

The Publishers regret that they can enter into no correspondence upon this matter.

THE ASSYRIANS

INTRODUCTION

For the greater part of the period from the end of the 10th century BC to the 7th century BC, the Ancient Near East was dominated by the dynamic military power of Assyria. At the zenith of its rule Assyria could lay claim to an empire that stretched from Egypt in the west to the borders of Iran in the east and encompassed for the first time in history, within the realm of a single imperial domain, the whole of the 'Fertile Crescent'. Yet within fifty years of its maximum expansion this empire had collapsed with remarkable rapidity — a consequence of over-extension and the material exhaustion attendant upon the defence of far-flung territories against external enemies, the suppression of internal revolts, and resurgent civil war as contenders for the imperial power vied for the throne of Assyria. In 612 BC the Assyrian capital of Nineveh was mercilessly sacked and laid waste by the combined arms of the Medes and the Babylonians. There were few who mourned Assyria's passing; and many among its former subjects would no doubt have lent their own voices to the paeon of triumph uttered by the Hebrew prophet Nahum on receipt of word of Nineveh's fall: *All who hear the news of you | clap their hands at your downfall. | For who has not felt | your unrelenting cruelty?* (Nahum Ch 3 vs 19).

Notwithstanding such sentiments, Assyria's empire provided a model for others to follow. The rise of the Neo-Babylonian Empire under Nebachadnezzar II and its Persian successor must be seen as conscious attempts by these powers, through their own rule over the Near East, to emulate Assyria's example and inherit her imperial mantle.

For many people it is in the pages of the Bible that acquaintance is first made with the Assyrians. Given the quite fundamental impact this book has had upon our own culture it is not surprising that the Assyrians have had less than a fair press, seen as they are through the distorting lens of a highly ethnocentric and theological view of history presented to us in the pages of the Old Testament. Such casual acquaintance has reduced Assyrian imperialism, in the minds of many, to a byword for cruelty and barbarous excess. A more balanced historical judgement, however, leads to the recognition that the civilisation of Assyria was of a high order and the legacy of its empire quite profound for the longer term cultural evolution of the Near East. Nevertheless, Assyrian imperialism was at its heart indistinguishable from a militarism embodied in an army which in terms of organisation, efficiency and effectiveness had no equal prior to the emergence of the Legions of Rome.

It had not always been so. Archaeological evidence testifies to a more peaceable situation at the start of the Second Millennium BC in which mercantile activities predominate in records of the period. However, by the end of the millennium there had been a marked shift towards a policy of aggressive military expansion, the motivations for which can be seen at work in the Middle Empire Period and carried over virtually unchanged into the period of the Neo-Assyrian Empire in the 10th to 7th centuries BC. As this period of Assyrian history is the subject of this book, a short examination of the roots of Assyrian militarism will be invaluable in understanding the imperial policies of the First Millennium BC.

THE LAND OF ASHUR

It was in the reign of Shamsi-Adad I (1813–1781 BC) that a political coherence can first be attributed to the geographical entity known as Assyria. This vigorous monarch gave form to the kingdom by establishing his state on the cities of Nineveh, Ashur and Arbil and on the rich agricultural wealth of the plains within this triangle of land. Such became the heartland of Assyria and was to remain so until the demise of the kingdom. A brief period of 'empire' under this monarch was to give way in the 15th century BC to the ignominy of vassaldom, as Assyria was incorporated within the realm of the kingdom of Mitanni. For the Assyrians the experience was seminal. In the absence of any natural boundaries to provide defensible frontiers, the productive grainlands of Assyria offered open and desirable access and rich plunder to any predatory nomads, mountain peoples or power who, fielding a superior military organisation, could impose its own will on the land of Ashur, as indeed happened with Mitanni. The only salvation in the face of such vulnerability lay in the conscious development and maintenance of an effective military that

would prosecute vigorous and, when necessary, ruthless offensive campaigns against those enemies who threatened the security of the Assyrian state.

Here in part lies the key to understanding Assyrian militarism. As the army, by the prowess of its arms, extended the defensive perimeter around the heartland it impinged upon the borders of more and more states, who in consequence saw their own security threatened by the growth of Assyrian power and thus in their turn became new enemies for the Assyrians to fight. This constant need to ensure military superiority honed the fighting skills of the Assyrian soldiery. Building on the organisation of the army, always the fundamental feature of its effectiveness, it forced the Assyrians to constantly modify their tactics and military technology in order to maintain their dominance on the battlefield. It was this unique flexibility, in part, which accounts for the success of Assyrian arms over the three centuries of the empire's existence.

Success in war has its material consequences; and the benefit to the Assyrian economy of vast quantities

The notion that the king was the agent of the god Ashur, and in expanding the domain of Assyria the monarch was executing the will of the god, is well illustrated here. The artist has depicted Ashurnasirpal at one with Ashur, shown here in stylised winged form, in the act of shooting the bow. It is not difficult to see how the Assyrians interpreted rebellion against the king as wilful opposition towards 'the lord Ashur'. (This and all other photographs so credited, courtesy of the Trustees of the British Museum)

of booty in the form of valuable metals, wood, live-stock, horses and indeed deported populations, combined with the benefits arising from control of the great trade routes, did much to raise the wealth of the state to a level above which the prosecution of war to maintain such wealth became not only desirable but essential. Thus the security needs of Assyria became indistinguishable, in time, from the material prosperity of the state itself.

Nevertheless, the quest for security and economic advantage are not enough to explain Assyrian militarism. In addition, there arose an ideological dimension to such expansionism in the conviction of the Assyrian kings that the god Ashur had laid upon them the task of unifying the world under his aegis. However, this conviction that Ashur had provided a theological warrant and therefore motivation for Assyrian expansion did not precede such, but rather evolved in parallel with the political and economic imperatives that were its true motors. Thus in all the wars fought by Assyria from the 13th century BC onwards it is possible to discern this theological im-

perative underpinning the expansionist drive. This could not be more eloquently expressed than in the words of Sargon II when he stated: 'Ashur, father of the gods, empowered me to depopulate and repopulate, to make broad the boundary of the land of Assyria'.

Such words also draw our attention to the central role of the king in the evolution of Assyria's power. As the embodiment of the state, the monarch was the arbiter of all matters to do with Assyria. Thus the king was at once the appointed agent of the god Ashur, the commander in chief of the army, and the determinant of domestic and foreign policy; thus the dynamism of the state was irreducibly linked to the personality of the king. Assyria's rise to empire in the Neo-Assyrian period is inextricably bound up with the dynamic and qualitative leadership of such kings as Ashurnasirpal II, Tiglath-Pileser III, Sargon II and Sennacherib. The corollary of this was, however, that ineffectual kings led to periods of dynastic weakness when Assyrian power underwent a temporary eclipse on the international scene.

Such an interval occurred following the murder of Tiglath-Pileser I in 1076 BC. The 'Middle' empire, which assumed its greatest extent under this militarily vigorous monarch, rapidly shrank after his death, so that by the mid-10th century it encompassed no more than those territories that formed the heartland of Assyria as constituted by Shamsi-Adad some eight centuries before. Successor kings were unable to withstand the onslaught on Assyrian territory by nomads referred to originally by Tiglath-Pileser as the 'Ahlamu-Aramaeans', thereafter foreshortened in the annals to Aramaeans. Originating in the Syrian desert lands, the initial assaults of the Aramaeans on Assyrian territory had been vigorously suppressed by the king in at least 28 punitive campaigns between 1100 BC and his death. Thereafter, and in the face of ineffectual resistance by his successors, the Aramaeans pushed en masse into the steppe of the Jazirah to the west of the Euphrates and also southwards into Babylonia. Here their descendants, known collectively as the Chaldeans, were to become a constant thorn in the side of the Assyrians and were also, in time, to become prime movers in the ultimate destruction of Assyria itself. Assyrian weakness was to last until 911 BC, when the accession of a vigorous new monarch to the throne of Ashur revived

Assyrian fortunes. With the nomadic tide spent and the wanderers themselves now settled in numerous small states dotted all over Syria, the Jazirah, the Euphrates bend and the Tigris valley, a resurgent Assyria under Adad-Nirari II (911–891 BC) initiated a process of expansion which was to culminate within 200 years in the largest empire the world had yet seen.

Kings of Assyria
Early Empire, approx. 1813–1755 BC
Period of Mitannian domination, approx. 15th C BC–mid-13th C BC
Middle Empire, approx. mid-13th C BC to the death of Tiglath-Pileser I, 1076 BC

Neo-Assyrian ('Late') Empire:

911–891 BC	Adad-Nirari II
890–884	Tukulti-Ninurta II
883–859	Ashurnasirpal II
858–824	Shalmaneser III
823–811	Shamsi-Adad V
(810–746 Period of weakness)	
745–727	Tiglath-Pileser III
726–722	Shalmaneser V
721–705	Sargon II
704–681	Sennacherib
680–669	Esarhaddon
668–627	Ashurbanipal
?627–624	Ashur-Etil-Ilani
?623–612	Sin-Shar-Ishkun
(612 Fall of Empire)	

THE NEO-ASSYRIAN EMPIRE

In the 26 years which covered the reigns of Adad-Nirari II and his son Tukulti-Ninurta II Assyrian arms were once more carried beyond the borders of the homeland. In numerous campaigns, which netted huge quantities of booty, the former king settled the southern border of Assyria with Babylon, forced the formal submission of the Aramaean cities of Kadmuh and Nisibin, and seized control of the Habur region. Tukulti-Ninurta consolidated the gains of his father,

bequeathing to his own son and successor a territory stretching from the River Habur in the west to the Zagros Mountains in the east and from Nisibin in the north to Samara in the south. Such was to provide the springboard for the ambitions of one of the most ruthless monarchs ever to occupy the throne of Ashur, and regarded by many as the true founder of the later empire.

Ashurnasirpal II

It was in 1845 that Austen Henry Layard first began excavations on the great Tell at Nimrud. Within 24 hours of the dig beginning the spades of his Arab workmen revealed the outlines of a number of rooms that were to prove to be part of the great palace of Ashurnasirpal II, and the identity of the remains contained by the Tell to be that of his capital city Kalhu (Calah). Of particular significance were the numerous alabaster bas-reliefs that were uncovered and which testify to the predominant role war played in the 24-year reign of Ashurnasirpal.

Certainly one noted authority has seen in his campaigns to the north, east and south a systematic strategy designed to create a ring of security around the heartland, prior to embarking upon adventures further afield. Such an assumption may go some way to explain the ruthlessness with which Ashurnasirpal suppressed rebellion against Assyrian rule in the earlier part of his reign. This is particularly true of his dealings with the Aramaean cities to the west of Assyria lying within lands bordered in the west by the Euphrates River, which for Ashurnasirpal and his son Shalmaneser III was regarded as 'Greater' Assyria's border. Within this boundary lands came under direct provincial rule — beyond that line states were to be brought within Assyria's sphere of interest by vassal relationships. Shortly after his accession he abandoned a campaign in the Upper Tigris valley upon receipt of the news of the revolt of the city of Suru on the Lower Habur, and force-marched his army over 200 miles, in mid-summer along dusty 'roads' and with the temperatures in the high 90s, to suppress the rebellion. The rapidity of the Assyrian advance had not been foreseen by the leaders of Suru, and the psychological impact of the army's unexpected appearance on the Lower Habur was a factor of which the king was clearly aware:

'To the city of Suru of Bit Halupe I drew near,

and the terror and splendour of Ashur, my lord, over-whelmed them. The chief and the elders of the city, to save their lives came forth into my presence and embraced my feet, saying: 'If it is thy pleasure, slay! If it is thy pleasure, let live! That which thy heart desi-reth, do!' ... In the valour of my heart and with the fury of my weapons I stormed the city. All the rebels they seized and delivered them up.'

In an oft-quoted section of the same text Ashur-nasirpal describes in some detail that the rebels were

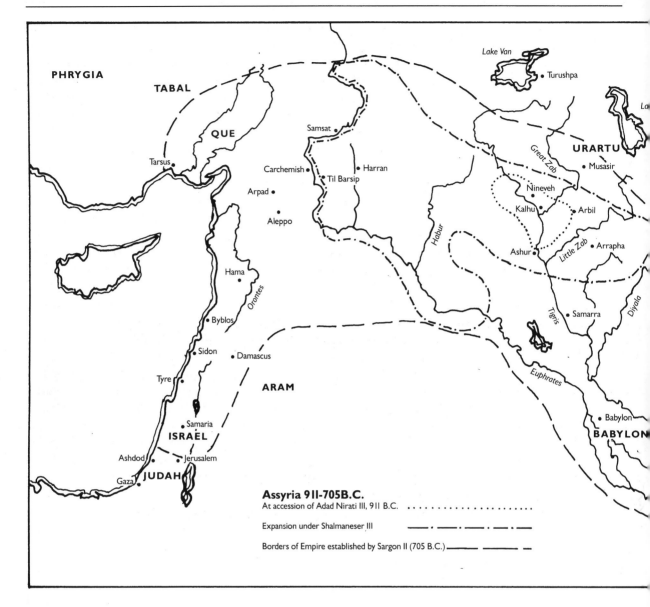

Assyria 911-705B.C.

At accession of Adad Nirati III, 911 B.C. ·

Expansion under Shalmaneser III ▬ · ▬ · ▬ · ▬ · ▬ · ▬

Borders of Empire established by Sargon II (705 B.C.) ▬▬ ▬▬ ▬▬ ▬

flayed, impaled, beheaded, burnt alive, had their eyes put out, noses, fingers and ears cut off, and so on. To modern eyes this grisly litany speaks of little more than sadistic excess. If, however, we are to place such behaviour in its historical context then a few pertinent observations need to be made. Whilst the relish with which Ashurnasirpal speaks of the punishments inflicted on his enemies finds no parallel in other royal annals, the rationale for such behaviour would have been understood and practised by earlier Assyrian monarchs and was to be a general feature of Assyrian policy until the fall of the empire. Indeed, the words of the prophet Nahum, quoted earlier, are an explicit reference to the deliberate and cultivated Assyrian policy of 'frightfulness'. Notwithstanding that such behaviour has become synonymous in the popular mind with commonplace Assyrian policy, it is quite apparent from the annals that the kings were highly selective in their use of it.

In most cases when atrocities do occur it is consequent upon a vassal state renouncing the solemn oath of fealty to the king and the god Ashur, the punishment being viewed as the rightful chastisement of a rebellious subject. Such a policy was practised specifically to dissuade opposition to Assyrian rule by setting an example from which other potential rebels should draw the appropriate conclusions. In wartime a few intentional and selective atrocities were engineered so

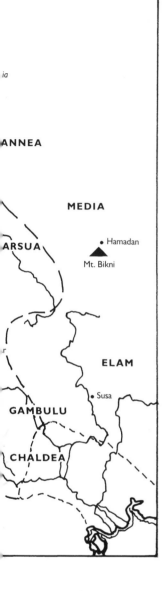

ia

ANNEA

MEDIA

ARSUA

• Hamadan

▲ Mt. Bikni

ELAM

• Susa

GAMBULU

CHALDEA

as to encourage the enemy to come to terms quickly, to avoid unnecessary battles. Thus the purpose of the policy was psychological, its principal object being to engender terror so as to convey a sense of power-lessness in the face of Assyrian arms.

Whilst Ashurnasirpal may indeed have been of a sadistic disposition, it is nevertheless reasonable to see his policy in the first part of his reign as a con-scious preparation for his later, ambitious push into

In an age when war was endemic it was universal practice to provide towns and cities with their own walls. It was therefore incumbent on any power with expansionist pretentions to conduct siege warfare on a sophisticated level. The Assyrian army throughout the period of the 'later' empire brought the art of the siege to a peak. Effective machines for destroying walls and other techniques such as mining were highly developed. The latter process is being employed in this relief, dating from the 9th century. The storming of the walls by scaling ladders is seen down to the time of Ashurbanipal. (British Museum)

Syria and the Mediterranean seaboard. Such a campaign, with extended lines of communications across lands filled by potentially hostile Aramaean settlements, could not be contemplated if armies far from home were in danger of being cut off by rebellions breaking out in their rear. If indeed that was the purpose then the policy would seem to have succeeded.

Having neutralised the powerful Aramaean kingdom of Bit Adini, Ashurnasirpal launched a campaign to the west, surprising the Neo-Hittite and Aramaean states of northern Syria and reaching Mount Lebanon and the Mediterranean in 877 BC. In the face of the overwhelming power of the Assyrian army many of the smaller states along the route of advance and in Phoenicia hastened to pay homage to the king, becoming vassals of Assyria. Whilst he made no attempt to annex these territories, the motives for his expedition become obvious from the account he gives of those who came to him:

'... The tribute of the sea coast — from the inhabitants of Tyre, Sidon, Byblos, Mahallata, Kaiza, Amurru and Arvad (Cyprus) which is an island in the sea, consisting of gold, silver, tin, copper, copper containers, linen garments with multi-coloured trimmings, large and small monkeys, ebony, boxwood, ivory from walrus tusk, a product of the sea, — this their tribute I received and they embraced my feet.'

With the cities of Phoenicia and northern Syria acknowledging Assyria's suzerainty, the principal benefit to the state of Ashurnasirpal's great raid was economic. In controlling the routes to the Mediterranean he secured an economic advantage for Assyria in assuring the supplies of raw materials vital to its burgeoning economy and to the war machine that made this possible. Along the trade routes into northern Syria flowed metals and horses from Anatolia, timber and luxury goods from the Lebanon and Phoenicia, in addition to the human resources of skilled manpower that Ashurnasirpal himself employed in the rebuilding of Kalhu. As Assyrian power increased so too did the strategic importance of Syria to its economy, with a consequent impact on the whole of Assyrian policy for the region. Nevertheless, incorporation of this vital strategic region as provinces within a formal imperial system would have to wait over a century until the accession of Tiglath-Pileser III. Notwithstanding, for the first time in two hundred years an Assyrian army had stood on the shores of the 'deep sea' and symbolically washed its swords in its waters. For the inhabitants of the region it cannot have been anything other than a fearful portent.

Shalmaneser III

The annals of Ashurnasirpal's successor tell us that he was at war for at least 31 out of the 35 years he reigned as king. In that time the tramp of Assyrian armies was heard in lands far beyond the confines established by his father. Nevertheless, while Shalmaneser III expended great energies on his martial expeditions, his reign ended with a sense of unrealised ambitions abroad and an Assyria plunged into civil war.

Ashurnasirpal II leads a parade, with arrows in his upraised right hand, symbolising the victory being celebrated. To his rear a trooper leads two replacement horses for the king's chariot; the horses themselves are decked out in ornate trappings. The king is shaded by a parasol carried by a eunuch; only the king's chariot could have a parasol, and like the polo crown worn by Ashurnasirpal it was a symbol of royal power. (British Museum)

Without doubt the greatest disappointment stemmed from his inability to bring to heel southern Syria despite five campaigns directed against the states in the region. In this he was frustrated by a coalition of small states led by Damascus, who sank their local differences in order to resist the greater threat posed to them by the might of Assyria. Shalmaneser's ambitions in Syria were writ large for all to see when, in the three years following his succession in 858 BC, the Assyrian army systematically reduced the former vassal state of Bit Adini. The capital city of Til Barsip on the Upper Euphrates occupied a strategic position, providing Shalmaneser with a bridgehead for operations against Syria. Bit Adini was annexed to become a province of Assyria, and over the next three years was transformed into a forward base for the Assyrian army in its planned operations against the west.

In 853 the Assyrian army crossed the Euphrates and, marching via the city of Aleppo, entered the plains of central Syria. Here they were confronted, at Qarqar on the Orontes, by a substantial coalition force of Syro–Palestinian states led by Adad-idri (Ben Hadad II) of Damascus and Irhuleni of Hamath. Included in the coalition total was infantry and chariotry supplied by Ahab of Israel. Although Shalmaneser claimed a great victory, it was in reality a check to the Assyrians. Three more western campaigns were also halted at Qarqar between 849 and 845 BC. Changing his route of advance in 841 BC, Shalmaneser laid unsuccessful siege to Damascus; but by way of compensation pushed into northern Palestine where, in addition to that of Tyre and Sidon, he received the 'Tribute of Iaua (Jehu), son of Omri', who, having recently become king of Israel following a bloody coup, now became an Assyrian vassal. One final attempt to conquer Damascus in 838 BC also ended in failure, after which Shalmaneser abandoned further attempts to bring southern Syria within the Assyrian sphere of influence.

Economic motives also account for his decision by 832 BC to move north beyond the Amanus Mountains in order to take under his direct control the iron ore sources of Cilicia. Iron was by this time crucial to

The Assyrian army was rarely defeated by rivers. This scene, dating from the 9th century but suitable for the whole 'later' empire, shows infantry issued with skins which they blew up and used to paddle across. Horses swam, guided by one of the swimmers. Heavier items of equipment such as chariots or siege engines were broken down and ferried across 'in boats made of skin', nowadays known as 'keleks'. (British Museum)

the Assyrian economy, particularly in the production of weapons for the army. Having no iron mines of her own and needing to import this strategic metal, it is conceivable that such a move may have been taken with a view to securing these supplies in the light of the possible threat to them from the emergent state of Urartu, north-east of Cilicia in the eastern Taurus Mountains. In an expedition to Urartu in 840 Shalmaneser claimed to have destroyed the royal cities of Sugunia and Arzashkun; but while the annals claim a great victory over the kingdom and report that the army returned weighed down with much booty, the rise of Urartu and the very real threat it was to pose to Assyria in the 8th century BC shows how little Shalmaneser's expedition actually achieved.

It had ever been the tradition of Assyrian monarchs to lead their armies into the Zagros Mountains on punitive expeditions against the mountain tribes that lived there. However, towards the end of his reign Shalmaneser consciously set forth to extend Assyrian influence beyond the Zagros into north-west Iran. The king, probably due to old age, delegated to his 'turtan' or field marshal Daian-Ashur the task of leading the Assyrian army. Advancing through Kurdistan, he crossed the mountains and came into contact for the first time with a number of new peoples. To the south of Lake Urmia lay the Manneans, whose importance as horse breeders was to make this area a major source of conflict between Assyria and Urartu in the next century. There were also the Medes who had settled in the vicinity of Hamadan; and the Persians (Parsua), who although established to the west and south-west of Lake Urmia were still engaged in the slow process of migration. Indeed, the Persians would not arrive in Fars, their final abode, for several generations, until after Ashurbanipal had finally destroyed the kingdom of Elam in 639 BC, whose territories the Persians took over and settled.

The meeting of these peoples and the Assyrians was significant and in the case of the Medes, momentous; for it was they, in conjunction with the Babylonians, who brought about the end of Assyria in the 7th century. From Shalmaneser onwards they are mentioned in the annals of every warlike king of

Assyria. At this time, however, the term Mede expresses an ethnic tribal concept rather than a coherent political entity; it was not until the late 8th century that the Medes formed a kingdom.

To the south of Assyria, Shalmaneser had intervened in the civil war that erupted in Babylonia in 851 BC. The established king Marduk-zakir-shumi faced a rebellion led by his brother and supported by the Aramaean tribes who were settled in the ancient land of Sumer. It was the first, but certainly not the last time that an Assyrian king took punitive action against these tribes now collectively known as the Chaldeans. In their repeated attempts to seize power in Babylon these Chaldeans would transform Assyria's dealings with her southern neighbour into a running sore, contributing in time to the draining of the empire's energies and playing no small part in its eventual downfall.

For all his military exertions Shalmaneser had not added significantly to the territories bequeathed to him by Ashurnasirpal, and when he died in 824 BC

Assyria had been at war with itself for four years. The accession of his son as Shamsi-Adad V marks the beginning of a period of decline in Assyrian power that was to last over 80 years.

THE ECLIPSE OF ASSYRIA

The resolution of the civil war in Assyria was only brought about by Shamsi-Adad V acquiring help from Babylon, which in its turn forced upon the king a humiliating treaty virtually reducing Assyria to vassaldom. In the subsequent recovery Shamsi-Adad took his revenge on the southern kingdom. Whilst he

This scene, from the black obelisk of Shalmaneser III, is believed to show the Israelite king 'Jehu, son of Omri' paying homage in approximately 841 BC. (British Museum)

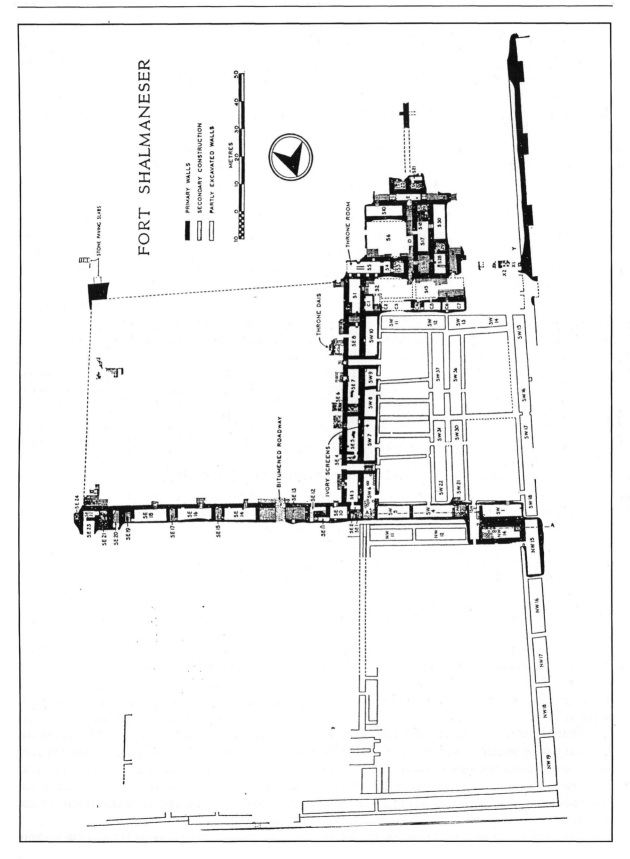

FORT SHALMANESER

PRIMARY WALLS
SECONDARY CONSTRUCTION
PARTLY EXCAVATED WALLS

STONE PAVING SLABS

METRES
0 10 20 30 40 50

THRONE ROOM

THRONE DAIS

BITUMENED ROADWAY

IVORY SCREENS

◀ The ground plan of the large 'ekal masharti' or arsenal built by Shalmaneser III in his capital city of Kalhu and given the name of 'Fort Shalmaneser' by its excavators. It was found to contain a large parade area, royal rooms, and rooms for the storage of food, weapons and booty. Cuneiform entries list the distribution of food to 'commander of the garrison company of war chariots', 'the Kassite and Assyrian slingers'. Such buildings established in many cities of the empire became the starting points for the army on campaign.

▶ *An Assyrian conical helmet from the late 8th century, made of iron and inlaid with bronze. (British Museum)*

also campaigned further afield in order to re-establish Assyrian supremacy over those states that had taken advantage of the civil war to loosen their vassal ties to Assyria, his early death marks a fairly rapid decline in Assyrian fortunes.

Attention has already been drawn to the synchronicity of Assyrian power and the qualitative leadership of the state provided when the throne of Ashur was occupied by a king of ability, dynamism and vision. The period between 811 and 745 BC is marked by a succession of weak monarchs unable to offer such leadership. During the period of the civil

war many states in northern Syria refrained from the payment of tribute; in this they were encouraged by Urartu, which in the period following the death of Shamsi-Adad moved rapidly to extend its influence over territories of vital strategic concern to the Assyrian state. A strong king would have ridden out the problem, as indeed Tiglath-Pileser III was to show; but the lack of such a one on the throne of Ashur at this time was to engender a situation in which unchecked Urartian expansion came perilously close to wrecking Assyria.

Paradoxically, the creation of Urartu can in many

ways be laid at the feet of Assyria. From the time of the 'Middle' empire strong Assyrian monarchs had led predatory expeditions into the Taurus Mountains without much difficulty, by virtue of the tribal organisation of the inhabitants; but by 840 BC a unified kingdom known as Urartu had emerged — indeed, it may well have been the raids of the Assyrians that forced the move to statehood. With its heartland around Lake Van and much of its territory in high mountains and buried under deep snows in the winter, nature had provided it with formidable natural defences. Evolving an administration and military organisation that drew heavily on the Assyrian model, this young and vigorous kingdom exploited to the full the decline of Assyria's power and prestige to expand into territories economically vital to her great southern neighbour.

In the reign of Argistis I (780-756 BC) Urartu annexed all the lands around Lake Urmia, interrupting the flow of luxury goods such as lapis lazuli which came to Assyria from Afghanistan through Iran. More importantly, it threatened the supply of horses from Mannea at a time when the Assyrian army was expanding its cavalry. The threat to the mobility of the army was of the highest order. To the west, Argistis asserted his authority over the small states of Asia Minor which had hitherto been vassals of Assyria. Under this king Urartu extended its control to within 20 miles of the northern Syrian state of Aleppo, virtually encroaching on the western borders of 'Greater' Assyria.

The deprivation of metals and other commodities and resources upon which the economy depended had profound repercussions in Assyria. Unrest emerged in the major cities of the kingdom. The impotence of the monarchy was further revealed in the manner in which it has been shown that provincial governors, in the absence of strong control from the centre, were ruling their territories as virtually independent fiefs. Whilst strong governors conducting

Elevated to the throne in 745 BC and taking the throne name of Tiglath-Pileser III, the former governor of Kalhu introduced reforms that transformed the Assyrian state. Before his death in 727 BC he had extended the boundaries of the empire to the widest extent yet seen. (British Museum)

campaigns in their own name clearly helped defend the frontiers, the situation was revealing of the extent to which, in the absence of strong royal leadership, Assyrian power was beginning to fragment. Matters came to a head in 745 BC when, in a bloody coup which eliminated the royal family, the governor of Kalhu was elevated to the kingship. In the new king, who took the throne name of Tiglath-Pileser III, the Assyrians not only had a military leader of the highest calibre but a visionary who realised that if Assyria was to survive she had no choice but to tread the path to empire.

THE GREAT REFORMER: TIGLATH-PILESER III

The rapidity with which Assyria recovered from her period of weakness to re-emerge as the dominant power in the Near East was a direct consequence of the fundamental reforms initiated by the new king. These were directed to the transformation of Assyria's state apparatus and imperial policies, in consequence liberating energies and resources that would determine the course of the empire during the last and greatest century of its existence.

Above all, there was the recognition that Assyria's interests could no longer be effectively served by the imperial model inherited from Ashurnasirpal and Shalmaneser III. Tiglath-Pileser expanded the boundaries of the empire by moving to annex and place under provincial rule former vassal states that had lain beyond the borders of 'Greater Assyria'. Where vassal states continued, their loyalty was carefully monitored by an appointed Assyrian official. If there was a failure to render tribute steps were rapidly taken — when such coincided with wider strategic interests — to place the territory under direct provincial rule.

In order to pacify newly acquired territories or punish recalcitrant vassals Tiglath-Pileser reintroduced, on a more formal basis, the policy of mass deportation of populations practised by earlier kings. However, this new policy was of an altogether different order of magnitude to that which had gone before. Examples include Iran, where in 744 BC he caused 65,000 persons to be deported; while two years later he moved 30,000 inhabitants from the area of Hamath in Syria and settled them far to the east in the Zagros Mountains. It was assumed that by deporting those in the population likely to sponsor rebellion, such political 'decapitation' would render the kingdom more amenable to their Assyrian overlords, breaking the will of those who were left and removing any desire to translate residual nationalist sentiment into further rebellion.

In parallel, he moved to ensure that his rule over the empire and its resources was absolute by breaking down the former provinces into smaller, more administratively effective units — within seven years of his accession over 80 such provinces existed. This in turn deprived the provincial 'grandees' of previous reigns of much of their power; to negate what remained he appointed as governors many 'sha reshe' or eunuchs who, having no descendants, owed their loyalty solely to the king. A small administrative tier, reporting to the monarch, was tasked with the inspection of the provinces and reporting back upon the performance and loyalty of the governors. The setting up of a communications system of riders that criss-crossed the empire allowed the rapid dissemination of orders from the king and delivery of reports from the provinces, which contained amongst other matters intelligence from operatives in the spy service set up by the king to monitor states beyond Assyria's borders.

The period of decline had also served notice that the army, as the great instrument of Assyrian aggrandisement, was no longer suited for the far wider demands made upon it by the constantly enlarging visions of imperial conquest it was required to serve. Tiglath-Pileser had perceived that a fundamental reform of the army, to parallel those of the imperial administration, was needed to ensure that Assyria's future military needs could be effectively served. What was entailed served only to demonstrate how the army of the Assyrian empire post 745 BC was a very different animal from that of even a hundred years before.

THE ASSYRIAN ARMY

One of the features of the pre-reform Assyrian army, in common with others in the ancient Near East, was that it was called up for war in the summer. The notion of a 'proper' campaigning season arose from the reality of the bulk of the army being drawn from the peasantry. The economy of Assyria was grounded on agriculture, and the peasants were needed to service the land and reap the harvest; this dictated their availability for military service. With the harvest collected in May, the call-up allowed the levies to be with the 'colours' by July. To these were added the chariotry of the nobility and those of the king's small standing army, in addition to support units. This seasonal aspect may well have been a determining factor in the strategy lying behind Ashurnasirpal's concept of 'Greater Assyria'. Operations within its borders, and 'raids' beyond, could be accommodated within the agricultural timetable and without major economic dislocation; but as the frequency and range of Assyrian military operations increased, as under Shalmaneser III, stresses developed. The greater frequency of wider-ranging campaigns, with their inevitable losses, made inroads into native manpower. Though Assyrian annals always indicate that such losses were low, attrition through sickness, the need to garrison fortresses and battle casualties inevitably had an impact. The finite availability of native Assyrian manpower and the constraints on its employment meant that the organisational basis of the army could no longer cope, by the time of Tiglath-Pileser, with the demands placed upon it.

The 'New Model Army'

His most important reform fundamentally changed the character of the army: moving from a seasonal conscripted native levy, he introduced a standing army available for service throughout the empire and at any time of the year. As important, the bulk of the manpower was to be provided by contingents raised from the provinces within the empire and augmented when needed by units provided by vassals, whose service was specified as part of the tribute levied by the king.

The 'kisir sharruti' or standing army acquired a diverse aspect, with many foreign troops incorporated alongside the native professional Assyrian units. These were issued with Assyrian uniforms and equipment and would therefore become virtually indistinguishable from one another within the ranks. Sargon II reported on the way he equipped and incorporated 50 Israelite chariot teams into the army. Other identified units include those from Hamath, Carchemish, and Greek settlements in Asia Minor. Aramaean units were heavily employed, with some tribes like the 'Itu'a' and 'Gurra' having a position somewhat analogous to that of the Gurkhas within the British Army. Nevertheless, it is clear that there was a very large, distinctively Assyrian element

within the 'kisir sharruti', serving principally in the cavalry and chariot arm.

Alongside the standing army served the 'qurubti sha shepe', or king's personal bodyguard, composed of cavalry, chariotry and infantry. On campaign the 'sa shepe' and 'kisir sharruti' would be reinforced by the 'sab sharri', levies raised by the provincial governors on the orders of the king. Their diverse origins and retention of native uniforms are readily observed in the bas-reliefs showing the wars of the later Assyrian kings. However, the availability of the 'kisir sharruti' meant that it was possible to employ just the standing army and the 'sa shepe' without recourse to the 'sab sharri' when needs allowed.

As supreme commander the king of Assyria very often led the army on campaign; but could, when necessary or desired, delegate the task to the senior of two 'field marshals' designated 'turtans' of the left and right, such titles arising from responsibility in battle for the appropriate wings of the army; it was the former who took precedence in the absence of the king. Apart from Daian-Ashur, other named turtans include Shamsu-ilu, again from the reign of Shalmaneser III, and Sin-ah-usur, brother to Sennacherib, who led many of the campaigns in that king's reign.

Horseflesh was of paramount importance to the Assyrian army. This scene illustrates the degree of care taken of the animals when in camp.

Having fed them, the groom is brushing the coat of one mount. Without large numbers of horses the Assyrian empire could not have existed.

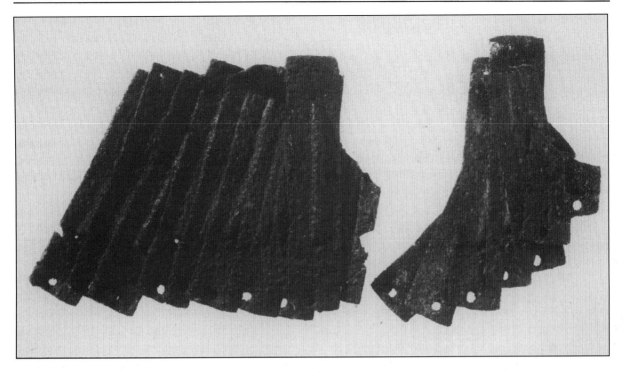

These samples of bronze lamellar armour show the holes through which stitching would have attached them to the leather underjacket. The reign of Tiglath-Pileser saw the introduction of such armour, which was widespread by the end of the empire. (British Museum)

Thereafter, and in common with most ancient Near Eastern armies, rank designation described responsibility, as in commander of '1000', '500' or '100'. However, there is much concerning the organisation of the army of which we are still uncertain.

The Arms of Service

Chariots

The heart of the army's offensive power lay in its chariot force. Over the three centuries of the 'Later' empire the chariot evolved from a two-man, two- and three-horse type during the time of Ashurnasirpal to the large four-man, four-horse chariot of the reign of Ashurbanipal. The lighter structure of the earlier chariots reflected their use in roles as diverse as reconnaissance, the carrying of despatches and combat; the presumption was that they would also need to serve in a wide variety of conditions. However, many of these roles were increasingly usurped by the cavalry as expertise in the use of horses was acquired, and the greater economy of this arm compared to that of the chariotry became apparent. By the final century of the empire the role of the chariot had been greatly reduced to that of a mobile firing platform for archers and a shock vehicle to be used in the charge against the enemy front. In the latter role the psychological and physical impact of such heavy machines bearing down upon and crashing into a line of enemy troops was beyond dispute, and in this situation the cavalry would be employed to support and then exploit the breakthrough made by these heavy vehicles. Nevertheless, the size and weight of these four-horse chariots was such that they were limited to involvement in battle on flat, open terrain, as at Halulue in 691 BC and Ulai in 655 BC.

Cavalry

The use of cavalry is first attested in the 9th century BC, in the reign of Tukulti-Ninurta II. Wall reliefs of Ashurnasirpal illustrate riders operating in pairs with one holding the reins of both horses in order to allow the other rider to fire his bow. Pairing of the cavalry is still evident in the reign of Tiglath-Pileser, although each rider operates as a self-contained unit controlling his own mount and armed with a long lance, used in a downward, overarm thrusting fashion. By the time of Ashurbanipal in the 7th century the horse warrior is an efficient and extremely well-armed and protected combat unit. Fabric body armour protected the mount to a limited degree from arrows and stab-

bing weapons when in close combat; the rider was also equipped with lamellar armour to the torso, and carried bow and lance. In the 7th century it is clear that the cavalry were a formidable and very important element of the army, having overshadowed that of the chariot arm.

The formative influence on the development of the cavalry arm was undoubtedly that of the Iranians. In their attempt to extend their control into and beyond the rough country of the Zagros the Assyrians were forced to adapt to the conditions of warfare imposed upon them by the Median and Persian horseman. Indeed, later Assyrian incursions into this region took the form of large raids on towns carried out solely by cavalry in order to capture, pillage and burn Iranian settlements and carry off booty. By the time of Tiglath-Pileser these raids thrust deeply into Iran, reaching as far as Mount Demavend near Tehran, testifying to Assyrian proficiency in the use of cavalry.

With cavalry units operating in numbers as large as 1,000 at a time, alone or in conjunction with chariots, it is clear that such large horse-dependent operations presupposed a highly efficient system for the supply of remounts. Indeed, without such the Assyrian Empire could not have existed. This was clearly evident to the Assyrian kings, who took a great deal of

In a conscious policy Tiglath-Pileser ordered that in the wake of rebellion thousands of people were to be deported all over the empire to discourage further revolt against Assyrian rule. (British Museum)

personal interest in the matter. Three main sources of supply are identifiable. The first was the capture of horses in raids specifically organised for that purpose. The second saw horses included as part of the tribute from vassal states. Most importantly, however, there was the system extant within the empire whereby high-ranking state officers known as 'musarkisu' were allocated on a provincial basis to oversee horse supply; significantly, they reported to the king and not to the governor of the province. Census returns for the animals were compiled, listing the type and where they were held. From the outlying provinces the horses were driven in large numbers to the 'ekal masharti' or military arsenals where they and their future operators were trained.

Infantry

Nevertheless, it was the infantry that always provided the bulk of the Assyrian army on campaign. Although known simply as 'zuk' or 'zuk shepi' it is clear that there were many different types of infantry, distinguished by their function as archers, lancers, slingers or shield bearers. Certainly the former were the largest in number. The bow was the Assyrian army's main offensive weapon, and it was amongst the infantry on any campaign that the bulk of such weapons were to be found. Archers were normally employed in groups, and from the time of Ashurnasirpal to the end of the empire in pairs, with the second man being a shield bearer. These shields varied in design throughout the late empire, although the largest employed and seen often in bas-reliefs from the reign of Tiglath-Pileser onward were taller than a man and curved at the top to protect the team from plunging arrow fire. They were made from thick plaited reeds, as were many other shields, and records show that special areas in rivers were set aside to grow reeds specifically for shield use. Bows of many types are referred to in Assyrian texts by names such as Assyrian, Akkadian and Cimmerian. While many employed were of a composite type, as frequently used by the cavalry and chariotry, more simple bows were also employed. Depending on the type the range of such weapons ran from 250 to 650 metres. Certainly facilities existed whereby new arrows could be manufac-

tured on campaign in order to replace the vast number expended by the whole army, although to begin with the supply trains would carry very large numbers of replacement arrows made prior to the beginning of the campaign. Used in conjunction with the archers were slingers, with the former preceding the latter in line abreast, providing high-angle fire to negate the enemy's shields, making them vulnerable to archery at low angles.

In the bas-reliefs from the early 7th century onwards it is possible to infer the section of the army in which a particular infantryman served by his equipment fit. Units of the 'zuku sa sheppe' or elite infantry units of the 'kisir sharruti' wear lamellar armour to the torso, and in the case of the latter this probably identifies their role as close order infantry. Such armour is lacking among the ordinary infantry and those native levies raised to serve in the army, whose personal protection is limited to a helmet and shield. It was the native levies that also provided the light skirmishers, the Chaldean and Aramaeans being particularly expert in this role. Tiglath-Pileser introduced the lance spear to the infantry; it was employed as a close-order thrusting weapon, giving its user a longer reach than with the sword or dagger.

Siege Warfare

In an age when virtually every large town or city was

A siege of what is thought to be a Median city by Tiglath-Pileser's army. Auxiliary troops scale the walls while archers covered by high wicker shields shoot at the battlements. Enemy captives hang impaled before the walls. Of particular note is the shape of the siege engine of a noticeably different design to that of the reign of Sennacherib — cf. Plate B. (British Museum)

protected by its own defensive walls the Assyrians could not have created an empire unless they had been remarkably proficient in siege warfare. However, siege warfare was something forced upon them rather than desired: their preference was always for open battle to force a rapid decision. It is in this context that the earlier reference to 'frightfulness' can perhaps best be understood as a tactic designed to encourage the enemy not to retire behind fortified walls. When sieges were undertaken the Assyrians tried to end them quickly. Even so, the very formidable defences of some major cities such as Babylon, Jerusalem, Samaria and Arpad led to sieges that lasted for years, tying down valuable Assyrian assets that could have usefully been employed elsewhere.

Sieges and assaults on city walls are a feature of the campaigns of all the Neo-Assyrian kings and are well illustrated on their bas-reliefs, clearly being the most complex of contemporary military undertakings. The variety of siege engines illustrated shows a willingness to experiment to identify the optimum design. Nevertheless it is significant that in extant reliefs of Ashurbanipal, the last great king of Assyria, there are sieges and assaults on cities which do not utilise siege engines.

The Assyrians were very methodical in their approach to sieges. The nature of the site and the fortifications dictated their strategy. Generally towns and cities were built either on large mounds or hills or with one or more walls covered by a river. In each case the sites were isolated by earthworks being thrown up around the city. In sealing off the site there was every intention of allowing starvation and disease to assist the besiegers. The method of dealing with cities sited on hills is well illustrated on the Lachish wall relief showing Sennacherib's siege in 701 BC.

A timber framework and earth rampart was thrown up against the wall. When it was completed, wheeled engines covered in large sheets of leather were pushed along a wide prepared track atop this ramp. Withering fire from massed archers and slingers, themselves protected by the ubiquitous shield bearers, was directed at the Judaean troops who lined the crenellated battlements. The intention was to allow the engine to come sufficiently close to allow the long spear-like ramming rod to close with the wall. To protect the engine from fire one of the crew had the task of dousing any flames using a large ladle. Ear-lier siege engines from the reign of Ashurnasirpal used wide blades to prize apart the stones or dried mud brick walls of the besieged city. When some of the walls faced out onto a river the Assyrians constructed siege towers which were floated into position to pour archery fire down onto the walls. Miners were also employed, and are frequently illustrated using iron and bronze tools to weaken the walls at vulnerable points in order to encourage a collapse so as to form a breach. When the moment was deemed right the walls would be breached and simultaneously assaulted by scaling ladders. The cost of such an assault was high — another reason why such sieges were avoided when possible. A mass grave at Lachish was found to contain the remains of 1,500 Assyrian soldiers.

On Campaign

Preparations for a new campaign would begin with the assembly of troops at the designated base for operations. In Assyria this would be at one of the 'ekal masharti' in Kalhu, Nineveh or Khorsabad. Within the empire this would include cities designated as such, as in the case of Til Barsip on the Upper Euphrates. Orders would have been received by the local governor to prepare for the campaign by amassing supplies of corn, oil and battle equipment. Other governors would have also received orders and called up their provincial levies. States allied to Assyria, who in fulfilment of their vassal oaths were to support the campaign, would be ordered to present their contingents of troops in good time. With the arrival of the king and units of the 'sa sheppe' and 'kisir sharruti' the army for the campaign would have been fully assembled. Such a process allowed the Assyrian kings to field very large armies when needs demanded: Shalmaneser III speaks of moving into Syria in 845 BC with 120,000 men under arms.

Under normal conditions the line of march assumed a fixed order. In the van came the standards of the gods and their attendant religious functionaries, followed by the king in his chariot. He would have been protected by units of the 'sa sheppe' with the support of the main body of army chariotry and cavalry to hand. To the rear of this large mobile force, available for immediate action should need demand, came the infantry of the regular army and the mass of provincial and vassal levies. Following behind the combat troops came the siege train, supply wagons and assorted camp followers. A number of accounts indicate that such a column could travel 30 miles a day in good going.

The annals do not provide us with any detailed accounts of the tactical methods employed by the Assyrian armies, but by inference the following disposition was probable on the battlefield. It is reasonable to assume that in order to retain the initiative the Assyrians would prefer to attack rather than be attacked. The main infantry battleline was composed of archers, slingers and lancers, flanked on either wing by chariots drawn up in concentration with cavalry. The latter either functioned as a shock force in their own right or supported the chariots by exploiting the breach created in the enemy lines. Long-range arrow and sling fire would begin the battle to unnerve the enemy and cause as many casualties as possible prior to launching the chariots and cavalry at the enemy wings. Here the intention was to allow the chariots to crash through creating a breach. The cavalry following through the breach would wheel and roll up the enemy line from behind. With the enemy line wavering, the infantry would advance all along the front in order to maintain and extend the gap in the enemy front. When it was not possible to fight set-piece battles the Assyrian army was remarkably flexible in its ability to adapt its tactics to suit difficult terrain and unconventional warfare.

The Campaigns of Tiglath-Pileser III

Within a very short time of his accession Tiglath-Pileser set about retrieving the dire military situation that had in no small measure been the cause of his elevation to the kingship. That he managed to do so, and also to expand the borders of the empire so quickly, was testimony to his remarkable abilities.

In his first year he took the Assyrian army south into Babylonia to aid King Nabu-nasir in suppressing the troublesome Chaldean tribes. Taking advantage

It was in the reign of Tiglath-Pileser that Assyrian cavalry were first issued with lamellar bronze corselets and lances. (British Museum)

of Babylonian weakness, he fixed the frontier between the two kingdoms on the River Diyala. Having dealt with the Chaldeans, he then made visits to the large cities of Babylonia, and claimed for himself the title of 'King of Sumer and Akkad' whilst actually allowing Nabu-nasir to retain the throne of Babylon; nevertheless, he appointed an Assyrian governor over the cities of Babylonia, in order to protect Assyrian interests there. This timely and vigorous intervention rendered Babylonia quiescent until 732 BC, allowing him to focus Assyrian resources and attention on the task of throwing back the advances Urartu had made in the previous 40 years and rectifying the economic ills to Assyria that had flowed from these Urartian gains.

To this end he began a systematic assault on the crescent of vassal states and influence that Urartu had constructed from Syria in the west through Urartu and eastwards into the Zagros and western Iran. His first target was the coalition of anti-Assyrian states led by Mati'-ilu of Arpad, who were all firmly in the Urartian camp. When the Assyrian army advanced into Syria in 743 BC Sardur II, king of Urartu, responded to the urgent requests of his vassals and brought his army to do battle with Tiglath-Pileser. In a major battle near Samsat in Kummuh on the Upper Euphrates the Assyrians inflicted a major defeat on the Urartians. Pursuing the retreating Urartian army, the Assyrians invaded Urartu itself, laying unsuccessful siege to the capital Turushpa on the shores of Lake Van.

In no position to render further assistance, Urartu abandoned her Syrian vassals to their fate. Advancing southwards on his return from Urartu, Tiglath-Pileser placed Arpad under siege; this was to last three years, and was only possible now there was a standing army. In the following year the Syrian states on the north-west coast and Phoenicia were pacified. The fall of Arpad in 741 BC and the southward march of the Assyrian army in 738 BC concentrated the minds of the kings of the states of southern Syria and Palestine wonderfully, and many made haste to embrace the 'feet of the king of Assyria' bearing gifts and tribute.

Tiglath-Pileser's motivation in wishing to control northern Syria has a familiar ring: he not only wanted to assure Assyrian trade with the west but also to gain control of commerce on the Eastern Mediterranean

Sargon II became king of Assyria in 722 BC, following a coup in which his elder brother (?) Shalmaneser V was deposed. He was a vigorous and able ruler who enlarged the boundaries of the empire bequeathed by Tiglath-Pileser (see map). He died in battle fighting the Cimmerians in Tabal in 705 BC. (A. Heinke)

seaboard. His movement into southern Syria-Palestine was to establish Assyrian control over the inter-regional trade along the seaboard and to redirect the flow of western commerce into Assyria proper. This naturally had an impact upon Egypt, who saw this region as vital to her own interests. When Tiglath-Pileser placed an embargo on the export of timber from the Lebanon to Egypt, a trade that had been carried on since at least the Third Millennium BC, determined efforts were made to undermine the Assyrian position in the region, and these account for subsequent rebellions by the states of Syro-Palestine and Assyrian campaigns in the years 734 and 733–732 BC. By that time he had reduced nearly all the states in the region to vassals and his armies had reached the 'Brook of Egypt' to the south of Gaza. Here he erected a stele to mark the southernmost boundary of the Assyrian empire.

In order to undermine the influence that Urartu had built up in the Zagros and Iran he launched a number of major campaigns designed to bring the whole of the region 'within the borders of Assyria'. He attacked the Medes in their lands to the north-west of Hamadan and deported some 65,000 of their number, whom he resettled along the Assyrian-Babylonian border on the Diyala River. The remains of an Assyrian provincial palace at Tepe Giyan are

◀ *Following his capture of Samaria Sargon tells us that he incorporated a number of Israelite units into the 'kisir sharutti'. The distinctive headdress of these soldiers has led to them being identified as Israelite. (British Museum)*

▶ In this relief found in his abandoned capital city of Dur-Sharrukin, Sargon II is seen speaking with a figure variously identified as one of his 'turtans' or the crown prince Sennacherib. If the latter, the crown prince was clearly given posts of major responsibility prior to Sargon's death including overseeing the northern border with Urartu. (British Museum)

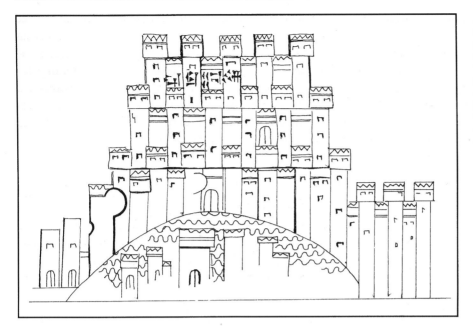

A highly stylised rendition of the type of Median defences encountered by Assyrian armies in their very frequent forays into the Zagros Mountains and beyond onto the Iranian plateau. The bas-relief depicts the town of Kishesim in north-west Iran.

testimony to the reach of Assyrian power in Iran during his reign.

In his 13th year he was once more involved in Babylon when he moved to assist the son of Nabunasir, who had been king of Babylon since his father's death two years before but whose throne had now been usurped by Ukin-zer, the chief of the Chaldean tribe of Bit Amukani. In the campaign that followed the Assyrian army entered Babylonia after crossing the Tigris in the vicinity of modern Baghdad. Although Babylon was taken, Ukin-zer fled to the southern Chaldean strongholds in the marshlands at the mouth of the Persian Gulf. The Assyrians followed him there and engaged in three years of bitter fighting, devastating his territories and those of other hostile Chaldean tribes before the area was pacified. Babylon was formally placed under Assyrian administrators, and in the 'New Year' festival Tiglath-Pileser took 'the hand of the god' and was declared king of Babylonia. Two years later, in 727 BC, he died, leaving Assyria more powerful than at any time in her history. He was without doubt a remarkable ruler, and probably the greatest king to sit on the throne of Ashur.

The Campaigns of Sargon II

The circumstances surrounding the accession of Sargon II in 722 BC are obscure; indeed, his relationship to Tiglath-Pileser III is itself uncertain, notwithstanding his own claim to have been the younger brother of Shalmaneser V, whom he replaced on the throne of Ashur. He may have been a usurper, as suggested by his deliberate and perhaps defensive choice of the throne name of Sargon, meaning 'true king'. Whatever the new king's origins he was the first of a dynasty of remarkable Assyrian kings who were to rule Assyria until the empire's demise and who are collectively known as 'the Sargonids'.

That his accession was not universally supported can be inferred from the fact that for the first year of his reign he undertook no campaigns beyond the borders of Assyria. As on previous occasions when internal dissension afflicted the homeland, trouble also broke out within the empire, with the first major danger emerging from Babylonia. Other upheavals occurred in Urartu and the kingdoms in the Armenian highlands, as well as Syria and Palestine. It was in 721 BC that Sargon personally led his army abroad for the first time, and southwards to contain the rebellion in Babylonia. This raises the interesting question of how we are to deal with the claim made in the later annals of Sargon, that he captured the Israelite capital of Samaria prior to his first Babylonian campaign. In these he states 'I besieged and conquered Samaria, led away as booty 27,290 inhabitants'. These deported Israelites, known thereafter to Jewish tradition as 'the lost tribes of Israel', were resettled throughout the empire including one group in Media. The siege of

Samaria had certainly been initiated by Shalmaneser V in 724 BC, and according to the biblical record (2 Kings Chap 17 v5) continued for three years. This accords with Assyrian sources and points to Samaria being captured in 722 BC, during the reign of Shalmaneser. It would therefore seem that Sargon appropriated this victory for himself although his suppression of the city in 721 BC, when it joined the anti-Assyrian coalition, led him to subsume both events in the one account.

The problem in Babylonia was not new; we have already seen how punitive measures had to be taken against the Chaldean tribes who made numerous and in some cases successful attempts to seize Babylon. Once more the same thing had occurred, although on this occasion Sargon II was faced with a wily and highly intelligent opponent in Merodoch-Baladan, who in the wake of the deposition of Shalmaneser V

had seized the throne of Babylon. Not resting on his laurels, he sought the support of the kingdom of Elam with whom he entered into alliance and began preparations for the inevitable Assyrian response. In proffering their support the Elamites were not acting out of altruism: they had many centuries of involvement in Babylonia, and saw in this situation a chance to muddy the waters to their own benefit.

The series of wall reliefs depicting the siege of Lachish by Sennacherib in 701 BC provide us with a highly detailed visual account of an Assyrian army at war at the very beginning of the 7th century BC. The detail was enhanced in the drawings made by Henry Layard in the 19th century. In this section the ramps thrown up against the walls of Lachish can clearly be seen. The city gates are the main target, being the obvious weak spot. Siege engines are used to effect a breach while behind them serried ranks of archers and slingers try to keep down the heads of the Judaean defenders. (British Museum)

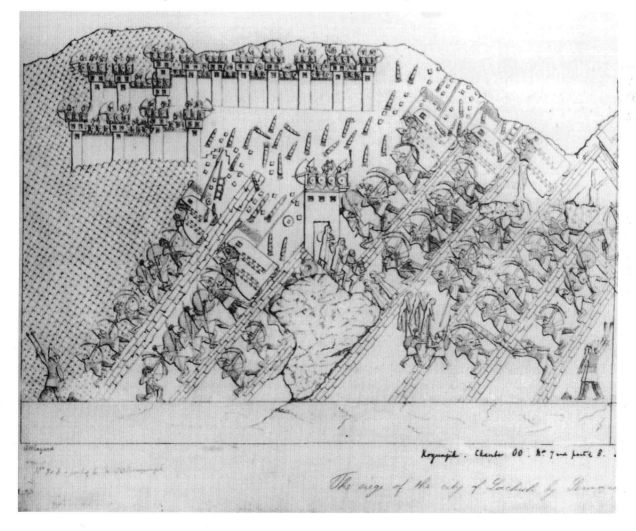

Advancing with his army along the right bank of the Tigris, Sargon was confronted by a very large Elamite force under their king at Der. In the subsequent battle Sargon claims to have 'smashed the forces of Humanigash'; however, the subsequent very rapid withdrawal of the Assyrian army and Sargon's abandonment of any further attempts to retake Babylon for ten years suggest that the Assyrians may have suffered a severe reverse. (The annals would not have recorded such an outcome. ...) What is clear is that the Elamites could field very large and well-equipped armies, and there is no reason to doubt that Sargon experienced a defeat. That an immediate riposte was not launched had much to do with the outbreak of rebellion in Syria, a matter of such convenient coincidence that some have seen the hand of Merodoch-Baladan at work in fomenting the revolt. Whether or not that was the case, Sargon was now required to turn his attention westwards.

The prime mover in the Syrian anti-Assyrian coalition was the ruler of Hamath, who with Egyptian connivance succeeded in drawing in Arpad, Damascus, Samaria and some of the cities of Phoenicia. Further to the south more overt Egyptian support led to Hanuna, king of Gaza, revoking his vassal oath and rebelling against Assyria. In a very rapid campaign Sargon moved south from his forward operating base at Til Barsip, first taking Arpad and then defeating the coalition army led by Hamath on the old battleground of Qarqar. Seizing Hamath, which was brought under direct provincial rule, he then moved south to capture Damascus, put down the rebellion raised by the rump of the inhabitants of Samaria, and finally appeared before the walls of Gaza with his army. An Egyptian expeditionary force was defeated in battle, and Hanuna captured (and promptly flayed) by Sargon. An attempt by Ashdod, again with Egyptian support, to create another coalition in 712 BC was rapidly suppressed, and thereafter Sargon had no further problems in Palestine.

In the two years 717–716 BC Sargon campaigned in northern Syria, placing the formerly independent city state of Carchemish under provincial rule. This action had been prompted by the fear of a conspiracy to control the trade routes through Cilicia and Anatolia by the king of Urartu and King Mita of Mushki, better known as Midas of Phrygia. This same concern prompted Sargon in 713 BC to annex the kingdom of Tabal after evidence that its king was attempting to ally himself to Phrygia and Urartu. In order to maintain pressure on Phrygia Sargon ordered Ashursharra-usur, the eunuch governor of the province of Que, to undertake military operations against Midas; under this Assyrian pressure, and with an eye to the

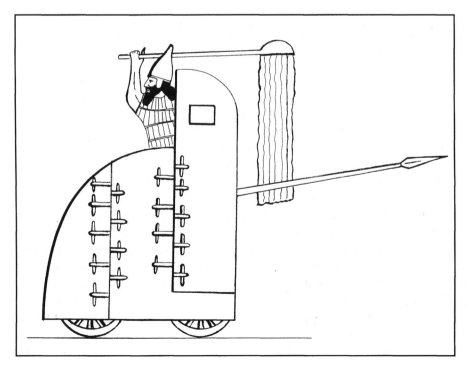

In this larger and more detailed view of one of Sennacherib's siege engines a number of features of note can be seen. The vehicle is clearly transportable, being made up of assembled parts. The wooden frame is covered with sheets of leather held together by toggles. To prevent the body being burnt one soldier uses a ladle to drop water. The long 'ram' is surmounted by a heavy iron 'spearhead'. Compare this engine with that shown in Plate B.

This bas-relief allows a look at the archers depicted at Lachish. All are protected by bronze lamellar corselets and the high, thick, reed shield held by the second men in the teams. Note the beardless archer at the front, no doubt a young 'apprentice' in arms. The appearance of nearly all infantry in the 'kisir sharutti' was to remain unchanged in essentials down to the fall of Assyria. (British Museum)

Cimmerian invasion of the Anatolian plateau, the latter bowed to the inevitable and in 709 came to terms with Sargon. With Urartu effectively neutralised by this time, the compact between Assyria and Phrygia further enhanced Assyrian domination and control of northern Syria.

Whilst Assyrian policy in northern Syria and southern Anatolia thwarted Urartian designs in that region, it was to the east in Media that the confrontation between these two powers was fought out. The region of Mannea, to the south of Lake Urmia in western Iran, has already been spoken of as being of vital importance to the Assyrians, particularly in the supply of horses. So, when Rusas I of Urartu succeeded in overthrowing the pro-Assyrian ruler of the region in 716 BC and installing as king a creature loyal to Urartu, Sargon was bound to act. He invaded Mannea, deposing the Urartian appointee and placing his own man on the throne. Matters had now come to such a pass that Sargon decided that only a major military enterprise would solve his problem with Urartu. Despatching his son, the crown prince Sennacherib, to the northern frontier, he tasked him with acquiring wide-ranging intelligence on Urartu. Spies were able to reveal that a major military disaster had befallen Rusas when his army was virtually destroyed by the nomadic Cimmerians. Such an opportunity could not be missed, and in 714 BC Sargon launched a major expedition against Urartu.

The Urartian Campaign

The remarkably detailed account of this expedition is preserved in a war report addressed to the god Ashur. In it Sargon described the passage of the Assyrian army through very difficult terrain of mountain passes, fast-flowing rivers and deep gorges. Sappers cleared paths to enable the chariots and cavalry to pass. A line of Urartian fortresses astride the obvious route of advance to Urartu, to the west of Lake Urmia, was avoided by the expedient of advancing around the eastern shore. Although the Assyrians were laying waste the countryside as they advanced it is clear that Sargon's principal objective was to bring the enemy army to battle: '. . . . because I had never yet come near Ursa (Rusas), the Armenian and the border of his wide land, nor poured out the blood of his warriors on the battlefield, I lifted my hands, praying that I might bring about his defeat in battle' . . . . The main Urartian army was finally located, drawn up across the valley bottom between two high mountains; clearly the Urartians believed they had chosen a position of great strength, and they challenged the Assyrians to give battle.

Sargon says of the condition of his own forces: 'The harassed troops of Ashur, who had come a long way, very weary slow to respond, who had crossed and re-crossed sheer mountains innumerable, of great trouble for ascent and descent, their morale turned mutinous. I could give no ease to their weariness, no water to quench their thirst; I could set up no camp, nor fix defences.' Eschewing the conventional deployment of his forces into line of battle Sargon chose to attack the Urartian army directly by assaulting one wing of the enemy deployment, in column, and straight off the line of march.

With himself to the fore in his battle chariot and supported by the 'qurubti sa sheppe' — the cavalry bodyguard — the Assyrian force crashed into the Urartian battle line, which broke in the face of this unexpected and ferocious assault. To his rear the bulk of the Assyrian forces, including archers and spearmen, then deployed rapidly to take advantage of the disintegration of the Urartian line. With the Assyrian cavalry in pursuit, the enemy forces began a rapid retreat. Such had been the ferocity of the assault that Rusas abandoned the battlefield, leaving behind him 230 members of the royal family, state officials, district governors, and many Urartian cavalry and other troops. In the subsequent panic the Urartian king abandoned his immensely strong capital of Turushpa on the false assumption that Sargon intended to march on it and seize the place.

Sargon's exact route of march is still the object of scholarly dispute. Nevertheless, what is clear is that as the Assyrian army began its march home it took every opportunity to lay waste Urartian territory in a classic scorched earth policy. In a subsidiary operation Sargon led a column of a thousand cavalry against the city of Musasir after despatching the bulk of his army back to Assyria. High in the mountains and seemingly secure against attack, it was unprepared for the Assyrian assault. As the home of the Urartian god Haldi, and the traditional site of the coronations of the kings of Urartu, it boasted an immense quantity of treasure, which Sargon took back to Assyria as booty.

Although Sargon's inscriptions indicate that his victory over Urartu was total, in reality the kingdom was able to maintain itself for the next century before being finally overcome by the invading Armenians. Nevertheless the damage Sargon had inflicted was grievous; and inasmuch as no further wars were recorded between Assyria and Urartu over the next century he had succeeded in removing the only real contender to Assyrian hegemony in the Near East.

In the wake of his victory over Urartu and with the rest of the empire quiescent, Sargon at last turned south to settle matters with Merodoch-Baladan in Babylon. With the full resources of his empire and army at his disposal it nevertheless took Sargon the three years between 710 and 707 BC to subdue the Chaldeans. In the northern Babylonian cities and in the capital itself the Assyrians were welcomed as liberators, the bulk of the population having become alienated by the behaviour of the Chaldeans. Indeed, this northern Babylonian support for Assyria was prevalent until the fall of the northern kingdom, with the inhabitants viewing strong, stable Assyrian rule as preferable to the nascent chaos associated with the Chaldeans. In 707 BC the Assyrian army invaded the tribal homeland of Merodoch-Baladan, who thought it prudent to flee to Elam, to plot, and fight another day. Sargon deported over 108,000 Chaldeans and Babylonians from the region in the hope of pacifying it.

The final military act of Sargon's reign, which was to lead to his own death in battle, followed the westward movement of the nomadic Cimmerians into Anatolia; there is some suggestion that they may have launched a raid into Assyria itself. Archaeological evidence suggests that Kalhu had suffered major destruction in the latter part of the 8th century which some have explained as the handiwork of the Cimmerians. Certainly Urartu, already severely weakened, was unable to withstand their passage. Perceiving their movements to be a threat to Assyrian

Chariot and infantry, 9th century BC (see text commentary for detailed caption)

A

Siege warfare, reign of Ashurnasirpal II, 9th c. BC (see text commentary for detailed caption)

B   3   2   1   2

**Early Assyrian cavalry, 9th century BC
(see text commentary for detailed caption)**

C

Tiglath-Pileser III on campaign, late 8th c. BC (see text commentary for detailed caption)

D

Assyrian and auxiliary infantry, late 8th c. BC
(see text commentary for detailed caption)

E

Sargonid cavalry, Urartu, 714 BC
(see text commentary for detailed caption)

F

Sennacherib at Lachish, 701 BC
(see text commentary for detailed caption)

G

Assyrian troops in Babylonia, early 7th c. BC
(see text commentary for detailed caption)

Royal guardsmen, 7th c. BC
(see text commentary for detailed caption)

I

Cavalry and infantry, 655 BC (see text commentary for detailed caption)

J

Heavy chariot, reign of Ashurbanipal, late 7th c. BC
(see text commentary for detailed caption)

K

interests in Tabal, Sargon moved north in 705 BC to do battle with them. It was here, perhaps in the battle he sought with them, that Sargon died. His body remained unburied, to be devoured by birds of prey. Sennacherib, Sargon's son and successor, who had quarrelled with his father, agreed with the high priests of Ashur in judging the ignominious manner of Sargon's passing to be the punishment of the gods.

The Campaigns of Sennacherib

This son of Sargon inherited the throne of Assyria having already demonstrated sound military and administrative ability. The dominating object of his reign was the rebuilding of Nineveh, which became the capital of Assyria and was to remain so until its destruction. Abroad, Sennacherib was to have great difficulties in his dealings with Babylon, which acquired the hallmarks of a virtually intractable problem for Assyria.

Notwithstanding the punitive expeditions of earlier Assyrian kings which had meted out very harsh and summary treatment to the Chaldean tribes, they erupted in rebellion once more in 703 BC when Merodoch-Baladan returned from exile. The term 'tribes' should not conceal what was undoubtedly a rich and powerful coalition of well-organised groups who saw in their own bid for the leadership of the southern kingdom the chance to resurrect Babylonian leadership in Mesopotamia and beyond. Inasmuch as this inevitably challenged Assyria, it was a project that in no way inhibited their repeated attempts to secure the kingship of Babylon for themselves.

With the active support of Elam, Merodoch-Baladan raised the whole Aramaean population of Babylonia, prompting the rapid despatch of Assyrian forces. After entering Babylon the Assyrian army was sent south into the tribal homelands, where they ravaged the region: 'In the course of my campaign, I besieged, I conquered and carried away the soil of ... the Bit Dakkuri tribe ... the Bit-Sa'alli tribe ... the Bit Amukkanni tribe ... the Bit Yakin tribe. I let my troops eat up the grain and dates in the palm groves, and their harvest in the plain. I tore down and demolished their towns and set light to them, and turned them into forgotten mounds.'

The Assyrians deported 208,000 people. No doubt believing this to have been an effective lesson Sennacherib appointed Assyrian officials to oversee Chaldea and enthroned a native Babylonian as puppet king; this at least preserved the outward appearance of Babylonian independence, and reflects Sennacherib's desire at this stage to treat Babylon gently. Matters further afield now demanded his attention.

A rebellion had broken out amongst the kingdoms of Philistia with Hezekiah of Judah as ringleader; not surprisingly, the conspiracy was supported by the Egyptians. That Hezekiah had been planning this rebellion for some time is supported by the preparations that had been carried out since Sargon II's last campaign in the region in 712 BC. The Bible is clear about the scale and range of Hezekiah's activities, none more impressive than the construction of the tunnel to bring water into Jerusalem from the pool of Siloam. All evidence points to a conscious decision to arm Judah and prepare it for war, with the presumption that Jerusalem would be put under siege.

Sennacherib moved rapidly southwards down the Mediterranean coast in 701 BC, making a show of force that brought many hostile cities 'to embrace his feet'. In Philistia only Ashkelon and Ekron held out. The former surrendered before an Egyptian army entered Philistia to give support to the rebels. At Eltekeh the combined forces of Egypt and Ekron were defeated, with the latter submitting shortly thereafter; Judah was now isolated. Turning inland, Sennacherib began a systematic devastation of the whole country, taking 46 'walled cities' including Lachish, whose siege we have discussed above (this same strategy was emulated by Nebuchadnezzar II of Babylon over a hundred years later). From there a column was despatched to Jerusalem, and there can be no doubt it was Sennacherib's intention to take the city; earthworks were thrown up around the walls, but the siege was not begun before an attempt had been made to secure a surrender. Rabshekah, the chief cup bearer to the king (who may well also have been Sennacherib's brother, the 'turtan' Sin-ah-user), appealed to the people of Jerusalem, offering reasons as to why Hezekiah and the city should surrender. Encouraged by the prophet Isaiah, Hezekiah refused to surrender, but eventually compromised with the Assyrians. 'Grasping the feet' of Sennacherib, Hezekiah paid a tribute as a token of his submission. The somewhat precipitate Assyrian withdrawal, interpreted in the Bible as a 'divine deliverance', is more rationally explained by the news that Babylon had erupted in rebellion once more and demanded the full attention of the king's army.

In a repetition of the campaign of two years earlier Babylon was once again taken and the Chaldean lands ravaged. Merodoch-Baladan fled to Elam, where he died shortly afterwards. Sennacherib determined to bring Babylon under full Assyrian control and appointed as its ruler his eldest son, the crown prince Ashur-nadim-shum (772). This change of heart reflects the ambivalence which many Assyrians felt towards Babylon and which lay at the heart of the northern power's dealings with her southern neighbour. Regarded as the second city of the empire, it was also viewed with reverence as a holy city and a great cultural centre. Such sentiments generated strong passions, and there were clearly pro- and anti-Babylon parties in the Assyrian court. Whilst initially sympathetic, the Assyrian king adopted an increasingly hostile policy as matters escalated.

Elamite support for the Chaldeans included extending sanctuary to leaders and tribes who even from exile conspired to foment trouble in Babylonia. This tacit support for Chaldean meddling in Babylonia prompted Sennacherib to strike directly at the problem by attacking Elam itself in 694 BC. Using

ships built and crewed by his Phoenician vassals, he launched a novel amphibious attack which achieved some success in the lagoons of the Persian Gulf and along the coastline of Elam.

In a totally unexpected move, the king of Elam seized the opportunity of the Assyrian army's absence in the south to strike at Babylon in a major counter-offensive through the province of Der. Occupying Babylon, they captured Sennacherib's son (who was probably later executed by the bowstring in the royal dungeons at Susa) and placed their own nominee on the throne. There now began an increasingly savage and spiralling cycle of conflict between Assyria and Elam. In 693 BC Sennacherib led an Assyrian army through Der to attack Elam, only to have the Chaldeans raise Babylonia in rebellion in his rear. Two years later a coalition army of Elamites, Chaldeans and Arabs blocked the advance of the Assyrian army southwards at Halulue on the Tigris. The battle that followed may well have been one of the bloodiest of ancient times. Sennacherib described the event in the following fashion:

'At the command of the god Ashur, the great Lord, I rushed upon the enemy like the approach of a hurricane ... I put them to rout and turned them back. I transfixed the troops of the enemy with javelins and arrows. Humban-undasha, the commander in chief of the king of Elam, together with his nobles ... I cut their throats like sheep ... My prancing steeds, trained to harness, plunged into their welling blood as into a river; the wheels of my battle chariot

▼ In this Layard drawing the slingers (top left), archers, and spearmen (bottom right) are presented together. On the ground, to the front of the slingers in the upper register, can be seen the sling stones used as ammunition. (British Museum)

were bespattered with blood and filth. I filled the plain with corpses of their warriors like herbage....' Although Sennacherib claimed the victory, it was both bloody and Pyrrhic, the army being so badly mauled that it would not take the field the next year.

With Elam unable, due to an internal dynastic squabble, to support the Chaldeans, Sennacherib launched a major drive through Babylonia culminating in the siege of Babylon. For nine months of star-vation and disease the city withstood the Assyrian army. Sennacherib, after nearly a decade of bloody intermittent warfare and with his eldest son and heir dead, had no room in his heart for pity. He gave over the great city to the sack: 'As a hurricane proceeds, I attacked it and like a storm, I overthrew it.... Its inhabitants, young and old, I did not spare and with their corpses I filled the city.... The town itself and its houses, from their foundations to their roofs I

▲ *Another Layard drawing showing in greater detail the scene from the Lachish relief of Sennacherib on his throne. Immediately above and in front of the king are the following words in cuneiform: 'Sennacherib, king of the world, king of Assyria, sat on a throne and the booty of Lachish passed before him'. To his rear are the infantry and cavalry of the 'sa sheppe' or royal bodyguard. (British Museum)*

devastated, I destroyed, by fire I overthrew.... In order that in future even the soil of its temples be forgotten, by water I ravaged it, I turned it into pastures ....' Not surprisingly, Babylonia was quiet for the rest of his reign, as indeed was almost the whole empire.

It was in January 681 BC that Sennacherib met his death, in a manner not at all uncommon for oriental potentates: he was murdered, whilst at prayer, by two of his sons.

The Campaigns of Esarhaddon

In the wake of his father's murder crown prince Esar-haddon had to employ military power to ensure the succession. The regicides, having failed in their attempt to secure the kingship for themselves and deny it to their youngest brother, escaped to Urartu.

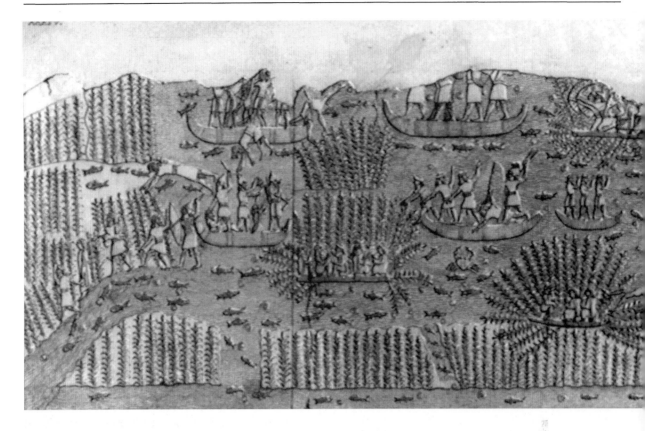

Entering Nineveh at the head of his army, Esarhaddon 'sat down happily on the throne of his father' in March 680 BC.

Belonging to the pro-Babylon party, Esarhaddon saw it not only as politically expedient but a religious duty to effect the reconstruction of the great city. Clever manipulation of religious texts allowed the new king to overturn the vengeful Sennacherib's injunction that Babylon should remain in ruins for 70 years. The level of destruction was such that the rebuilding took the whole of his reign, and indeed Esarhaddon did not live to see the return of Marduk and the other gods from exile in Assyria to be reinstated in their temples in Babylon. Much of the booty from his Egyptian campaigns was employed to finance this massive rebuilding project. Such efforts did not go unrewarded, for apart from an abortive Chaldean rebellion in 680 Babylon was quiescent for the rest of his reign.

Ever a superstitious man, there is evidence that Esarhaddon made repeated supplication to the gods and sought omens to guide his hand in dealing with the emergent problems to the west and north-west of Assyria. Cimmerian pressure was now felt in an arc from Anatolia through to Iran, and this was compounded by the appearance, at the beginning of his reign, of another nomadic group, the Scythians (see *Men-at-Arms* 137). A joint attack by these nomads on Tabal in 679 BC was quickly defeated by Esarhaddon; deflected westwards, the Cimmerians and Scythians fell upon and destroyed Phrygia. In north-western Iran the Scythians displaced the Cimmerians and established themselves in the strategically important country of the Mannai to the south and south-east of Lake Urmia. The concern about horse supply was compounded by another major development further to the south-east, where on the vast Iranian plateau the formerly disparate Median tribes had attained some semblance of political unity.

Assyrian interests were certainly not served by such developments and several raids, including a major offensive in 676 BC, were designed to exploit residual tribal rivalries and choke off this nascent political unity. It is a measure of Esarhaddon's perception of the growing power of the Medians that they figured so prominently in the treaty designed to ensure the smooth succession of his sons, Ashurbanipal and Shamash-shum-ukin, at his death. The mar-

◀ *The vicious wars fought by Assyrians and Chaldeans saw the former resort to fighting in the marshes and lagoons of southern Babylonia to root out the Aramaean rebels. In this Layard drawing from a bas-relief Assyrian soldiers use reed boats to attack the marsh dwellers. (British Museum)*

▼ *One of Sennacherib's most imaginative campaigns was an amphibious descent on the coast of Elam in 694 BC using ships built and crewed by his Phoenician vassals. (British Museum)*

riage of one of Esarhaddon's daughters to the Scythian chief Bartatua was as much an attempt to keep the peace with them as it was to employ the Scythians as a counterbalance to growing Median power on the Iranian plateau.

It was in the west, however, that he undertook the great military enterprise of his reign. It is possible that Esarhaddon came to the throne convinced that in order to completely secure Assyria's control over the sea and land trade of the Mediterranean seaboard he would have to eliminate Egyptian attempts to undermine their position there, and that such an end could only be secured by invading Egypt itself. Egyptian meddling in Palestinian and Syrian affairs, which culminated in the revolt of Tyre, a previously loyal vassal, was symptomatic of this. As early as 679 BC an Assyrian army had captured Arzani, the last major settlement before the borders of Egypt itself. Preparations now began for what was to be the greatest military expedition ever undertaken by Assyria. The invasion marks a major reversal of Assyrian policy towards Egypt, as previously the intention had been to keep Egypt contained within her borders. Not-

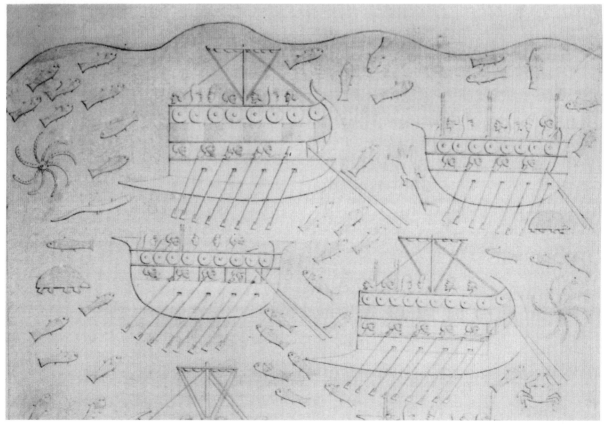

withstanding the bluster with which Esarhaddon later declared himself 'King of Egypt', it would seem that there was never any intention of permanently occupying the country. The timing of the invasion would seem to have arisen from an awareness that the ruling Ethiopian dynasty was not popular; fragmented politics and divided loyalties within the country seemed to offer the Assyrians an opportunity to use their military power to displace Taharqa and impose a native regime sympathetic to themselves.

The real challenge of the campaign, however, lay in the immensity of the resources necessary to sustain what was clearly a very large Assyrian army at the end of a line of supply and communications over a thousand miles long. Although the royal inscriptions do not detail matters such as manpower strength for the campaign, it is reasonable to assume that it involved a substantial amount of the total manpower serving in the Assyrian army, as a significant expansion in the size of the 'kisir sharutti' occurred at this time. With Assyria already defending a frontier of over 1,200 miles in the north and east, this major expansion in military commitment may have been regarded by some within the kingdom as a hostage to fortune; an enigmatic text dating to 670 BC speaks of Esarhaddon 'putting many of his nobles to the sword', a possible reference to the suppression of opposition that may have arisen over his Egyptian policy. (Indeed, it has been suggested that his death may have been due less to his inherent bad health than to poison.)

An initial attack on Egypt in 675/674 BC, about which Assyrian records are silent, was defeated at the frontier post of Sile. Three years later Esarhaddon

had greater success. Crossing the Sinai with the aid of his Arab vassals the Assyrian army entered Egypt, and in a number of battles ousted Taharqa, capturing Memphis on 11 July 671 BC. In his annals the king states: '… I fought daily, without interruption against Taharqa, King of Egypt and Ethiopia, the one accursed by all the great gods. Five times I hit him with the point of my arrows inflicting wounds from which he should not recover, and then I laid siege to Memphis, his royal residence, and conquered it in half a day by means of mines, breaches and assault ladders.' So precipitate was the Pharaoh's departure that he abandoned his son, harem and treasury. Esarhaddon settled an Assyrian administration on Lower Egypt, ruling through a number of native appointees.

The king returned to Assyria, where within two years he learnt of the return of Taharqa, who had moved his army back into the Delta region. It was while on his way back to Egypt at the head of a large army that Esarhaddon met his end, whether by fair means or foul.

The Campaigns of Ashurbanipal

The elaborate succession arrangements organised by Esarhaddon, whereby Ashurbanipal, his chosen successor, became king of Assyria and Shamash-shum-ukin, another son, king of Babylon passed off peaceably. While the reign of Ashurbanipal was to see the Assyrian Empire reach its zenith, his later years also saw a reversal of fortune and the onset of major, if not terminal, decline.

His first military decision as king was to confirm his father's plans, ordering the army despatched by Esarhaddon to continue its march on Egypt. It rapidly suppressed Tarharqa's attempts to raise Lower Egypt; and more local princes were selected to rule,

A bronze and copper mould used by Assyrian foot and horse archers for the casting of arrowheads of the 'Scythian' type: 7th century BC. (British Museum)

with the assistance of Assyrian garrisons. However, when evidence of their collusion with the Pharaoh emerged all were deported to Nineveh save one, who was returned to Egypt and confirmed as ruler of the Delta region, taking the name of Necho I. The accession of a vigorous new Pharaoh saw an Egyptian army sweep northwards into Lower Egypt and lay siege to the Assyrian garrison in Memphis. Ashurbanipal reacted promptly, personally leading his army against Egypt in 663 BC. Having relieved Memphis, the Assyrian army then marched south and sacked Thebes 'as if by a floodstorm'. The news of Egyptian resurgence had prompted Tyre and Arvad to rebel, however, and large forces were left to conduct siege operations against them.

A major campaign was launched against Media some time between 665 and 655 BC. Ashurbanipal was now confronted by a unified kingdom that had reduced Parsua to vassaldom. An abortive attempt by the Median king Khshathrita to attack Nineveh in 653 BC was defeated when the Scythians, allied to Assyria, took the Median army in the rear. With their

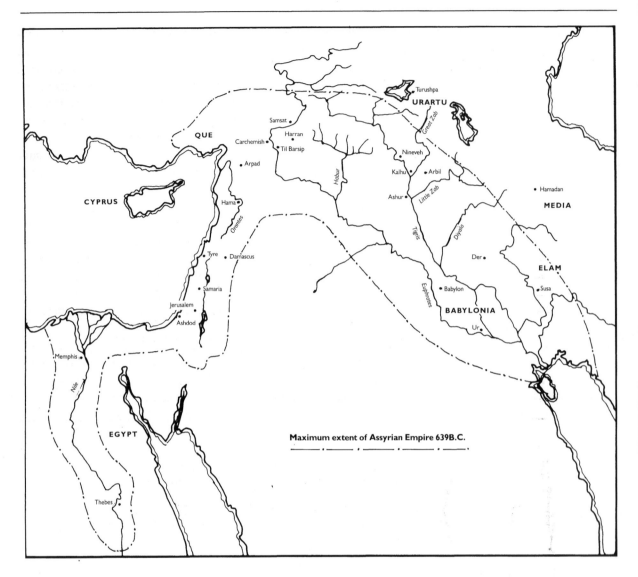

Maximum extent of Assyrian Empire 639B.C.

enemy's army destroyed and king dead, the Scythians moved against Media itself and for the next 28 years ruled over them. While Media was temporarily neutralised and the Scythians counted as allies, Ashurbanipal was able to use his military power to deal with the problem of Elam and Babylon.

It was while the Assyrian army was in Egypt in 665 BC that the first Elamite attack on Babylon since the reign of Sennacherib occurred. Although it was quickly repelled, relations between the two kingdoms deteriorated. Ten years later the coincidence of a major Egyptian offensive designed to drive out the Assyrians with that of an opportunist Elamite campaign into Babylonia in 655 BC served to demonstrate how stretched Assyria's military resources were becoming. Assyrian power in Egypt was at best

ephemeral and dependent on the availability of strong military force. The onset of these major strategic difficulties and the recognition of the practical limits of Assyrian military power prompted Ashurbanipal to act. By 651 BC, unable to spare the main field army to suppress the resurgence of Egypt and simultaneously deal with Elam and a major civil war in Babylon, Ashurbanipal abandoned the kingdom of the 'two lands'.

The aforementioned Elamite campaign in 655 BC was directed against Babylonia and assisted by the Aramaean tribe of the Gambulu. This was checked when Ashurbanipal moved south against the invaders with a very large army, pushing into the province of Der. Advancing rapidly, the Assyrians forced an Elamite retreat and with Ashurbanipal's army snapping at his heels Teumann, the Elamite king, determined to make a stand near the River Ulai in the plain of Susa. A major battle took place, with the Assyrians assaulting the Elamite defensive positions. In very heavy fighting the Elamites were pushed back against the river by Assyrian cavalry and infantry, many thousands being drowned as they tried to cross the Ulai. Teumann was caught while attempting to escape in his chariot, and beheaded. In the wake of his victory it is clear that Ashurbanipal did not have the strength to annex the kingdom, merely dividing the land between two Elamite vassal rulers (and thereby storing up future trouble for himself).

This arrived almost immediately when in 652 BC, with Elamite connivance, Shamash-shum-ukin raised the banner of revolt in Babylon, initiating a vicious civil war. A conspiracy engineered by the king of Babylon, and involving many of the subject states of the empire in addition to Egypt and Elam, came to nought. Once more the Assyrian army found itself treading the same path trodden by its predecessors in centuries gone by. Moving southwards into the marshlands of the delta after having placed Babylon

and Borsippa under siege, the troops of Ashur ravaged the lands of the Chaldeans for the last time. Hope that support would be forthcoming from Elam receded when news of the outbreak of civil war there reached the rebels: they were now totally isolated. Even so, 'the war went on and there were perpetual battles'. In 648 BC, with the population of Babylon reduced to cannibalism, Shamash-shum-ukin surrendered the city, but for himself chose death in the flames of his palace.

In the following years Ashurbanipal moved to punish those who had supported his brother. A desert campaign against the Arabs brought further victories, but it was only in 639 BC that Ashurbanipal decided to initiate the 'final solution' to the Elamite problem. By this time Elam was in terminal decline, rent by civil war and dynastic strife, and with its northern

In 663 BC Ashurbanipal led his army into Egypt and sacked the city of Thebes; he claims it fell in less than a day. Certainly it lacked the sophisticated defences of the cities in Mesopotamia, Syria and Palestine. Massive booty was taken from Thebes back to Nineveh. (British Museum)

borders already invaded by the Persians, who were shortly to make the region their final home. In the last great offensive war waged by the Assyrian army Elam was systematically ravaged in a massive campaign designed to destroy absolutely any vestige of opposition to Assyria. In triumphal tones Ashurbanipal tells us that:

'... For a distance of a month and twenty-five days' journey I devastated the provinces of Elam. Salt and sihlu I scattered over them. . . . The dust of Susa, Madaktu, Haltemash and the rest of the cities I gathered together and took to Assyria ... The noise of people, the tread of cattle and sheep, the glad shouts of rejoicing, I banished from its fields. Wild asses, gazelles and all kinds of beasts of the plain I caused to lie down among them, as if at home.'

And yet, having described Assyrian success seemingly at its apogee, the inscriptions of Ashurbanipal fall silent for the rest of his reign. It is in the silence of those years that is written the fate of the Assyrian empire.

'NINEVEH IS LAID WASTE'

Only tantalizing hints survive to inform us about the events of the last 12 years of Ashurbanipal's reign. Such as they are, they speak of major military disaster and civil strife within Assyria itself.

The most telling testimony of what would seem to have been a very rapid decline in the effectiveness of the Assyrian army lies in the virtual impunity with which the nomadic Scythians raided Assyria and rode roughshod over the lands of the western empire. Eusebius, a later Christian writer, speaks of them having reached Palestine by 633 BC and being bribed by the Egyptian Pharaoh to halt their advance. The latter observation also indicates Assyrian withdrawal from that region. Herodotus claims Ashdod was taken by storm by the Egyptians in about 635 BC. It does not follow that this speaks of Egyptian-Assyrian hostility; indeed, there is evidence to suggest that a rapprochement led to Egyptian power replacing Assyrian in the region, as perhaps the latter withdrew military assets and reduced military commitments in order to face greater problems nearer home. Although one of Ashurbanipal's successors may have conducted a campaign west of the Euphrates as late as

At the Ulai River in 655 BC the Assyrians inflicted a major defeat on the Elamites. In a combined arms assault the Elamite lines were carried and they were forced back to the river. Careful study of this drawing from a bas-relief will reveal much detail, including the execution of prisoners, and the appearance of auxiliary troops in Ashurbanipal's army. (British Museum)

622 BC, it would seem that by 627, the year of his death, Assyria was no longer able to operate in force in the eastern Mediterranean seaboard.

Within a year of Ashurbanipal's demise his sons Ashur-etil-ilani and Sin-shar-ishkun were at war with each other. The victory of the new Assyrian king over his brother Ashur-etil-ilani was overshadowed by the last and successful attempt by the Chaldeans to claim the throne of Babylon. In 626 BC Nabopolassar, the 'son of a nobody', seized the throne and was crowned king. The Assyrians still had powerful forces in Babylonia, in addition to political support, particularly amongst the cities in the north of the region; however, their inability to oust Nabopolassar seems indicative that the army was but a pale shadow of its former strength. Intermittent warfare continued until 616 BC, when the Babylonian Chronicle, our chief source of information in this period, tells us that Nabopolassar took the offensive against Assyria. Indeed it is possible, notwithstanding the growing strength of Babylon, that this unresolved state of conflict may have continued for some time. However, it was the

In this lively bas-relief the Assyrian artist has captured the distinctive appearance of the Elamite soldiers who for the greater part of the 7th century provided the Assyrian army with their most implacable enemies.

As is seen in the relief, archers constituted the great strength of the Elamite army. On occasions these troops bested those of Assyria. The chariot design and the mules pulling it are noteworthy. (British Museum)

appearance on the scene of the Medes, seemingly acting independently of Nabopolassar, that finally sealed Assyria's fate.

Having liberated themselves from the Scythian yoke, the Medes, under their king Cyaxares, reorganised their army on a more conventional 'all-arms' basis, taking the army of Assyria as their model. In 615 BC the Medes suddenly invaded Assyria, seizing the eastern city of Arrapha; and the following year they marched directly on the capital itself. Frustrated by the massive defences of Nineveh, they turned south and fell upon Ashur in substitute, where having breached the walls they: '... inflicted a terrible massacre upon the greater part of the people, plundering the city and carrying off prisoners from it'. An

alliance between Babylon and Media was concluded and sealed beneath its destroyed walls when Amytis, the daughter of Cyaxares, was betrothed to Nebuchadnezzar, the son of Nabopolassar. Two years later, notwithstanding Egyptian troops now serving alongside those of Ashur, the combined armies of the allies put Nineveh itself to the siege.

Recent archaeological discoveries at Nineveh confirm the accuracy of the description of the strategy employed by the Medes and Babylonians in taking the city which is to be found in the Book of Nahum in the Bible. A combination of assaults on the walls, the possible diversion of the course of the River Khosr to wash away defences, and widely separated attacks against the northern and southern gates of Nineveh were employed. Excavations at the south-western Halzi Gate and the northern Adad Gate all show how their widths of seven metres had been narrowed to two to render them more defensible. The discovery of skeletons with evidence of parry blows to the forearms and thrust wounds to the chest is testimony to the savagery of the final assault when it came.

It was left to some anonymous Babylonian scribe, taking up his reed 'qan tuppi' to inscribe the words of Nabopolassar on a tablet of clay, to write the epitaph of an empire that had cast its shadow over the whole Near East for three centuries:

'... I slaughtered the lands of Subarum (Assyria), I turned the hostile land into heaps and ruins. . . . The Assyrian, who since distant days had ruled over all the peoples, and with his heavy yoke had brought injury to the people of the Land, his feet from Akkad I turned back, his yoke I threw off'.

Even as the Median and Babylonian armies returned home replete with plunder and prisoners for the slave markets, the dust was already blowing into the desolate streets and the blackened remains of palaces and temples. Within two hundred years, as Xenophon was to show, even the identity of Nineveh was to be lost. In her destruction there were none to be found mourning her passing.

THE PLATES

A: Chariot and infantry, 9th century BC

The chariot design illustrated in this plate is characteristic of those that served with the armies of Ashurnasirpal II and Shalmaneser III. They are frequently shown with three horses, where the third horse would appear to be an outrigger and not attached by harness to the main chariot pole; however, the artistic rendition of the vehicle charging also allows for a three-horse chariot. In this particular case the chariot has only two horses minus the outrigger, as attested in the visual sources.

The long bronze scale 'sariam' worn by the archer illustrates the strong residual influence of Mitannian design. The distinctive Assyrian conical helmet is at this stage made of bronze rather than iron. The principal weapon employed was the compound bow. These charioteers are frequently illustrated 'doubling up' in sieges as dismounted miners.

Alongside are two infantrymen, both native Assyrians. They are élite troops belonging to the small standing army, and this is indicated by their uniform issue, which is superior in quality to that worn by the native levies. In the case of the spearman a variety of shields have been identified, running from the woven reed to the bronze-embossed circular wooden type. An officer is identified as such by the insignia on his helmet and the mace which he carries as a weapon as well as the symbol of his authority.

B: Seige warfare, reign of Ashurnasirpal II, 9th century BC

The centrepieces of this plate are the large siege machines. Whereas the artist in the original depicted the 'ram' and the tower as if they were one machine, in reality they were two, set alongside each other. The nearer is a six-wheeled engine with a total length of between four and six metres and about three metres in height. The 'ram' was suspended from a heavy rope attached to a crosspiece within the heavy iron

In this rather prosaic scene Ashurbanipal is shown dining with his first wife surrounded by fan bearers and musicians. To the left, hung from a tree by a ring, is the head of Teumann, the Elamite king. Ashurbanipal is reputed to have both slashed it and spat upon it when it was received at Nineveh. (British Museum)

cap. Although designated a 'ram' this is an inaccurate description of its function, as there is a wide iron blade which was inserted between the stones in the city walls and moved from side to side to prise them apart. Behind the 'ram' is a tall siege tower that allows archers to provide covering fire for the former machine. Ashurnasirpal (1) personally took part in sieges, as depicted here. Wearing the polo crown of Assyria, he joined his troops in shooting arrows at the defenders on the battlements. Clearly indicated are his shield bearers (2). To his immediate rear is a 'sa reshe' or eunuch of the imperial household (3). Other Assyrian troops visible are archer levies (4); their dress is far simpler than the élite infantry in Plate A, and is characteristic of the bulk of the troops in the army. Of note are the crenellated battlements, a characteristic feature of all military architecture in the Near East in the Neo-Assyrian period.

A close-up of an auxiliary soldier from the reign of Ashurbanipal. Apart from his crested helmet his only protection is his large shield and the bronze 'irtu' disc on his chest. (British Museum)

C: Early Assyrian cavalry, 9th century BC

Cavalry first served in the Assyrian army under Tukulti-Ninurta III. Those illustrated date from the reign of Ashurnasirpal, and show how the cavalry still employed the 'donkey seat' when riding the horse. Tactical employment in this period shows how, by riding in pairs, they were envisaged as 'charioteers without their chariot'. As in the chariot, the warrior (C1) is the superior soldier, as can be seen by his dress. The 'squire' (C2) wears a simple iron skullcap, which in the reign of Shalmaneser III had been replaced with an iron conical helmet of the type worn by the archer. Bronze bands let into or standing proud of the helmet were an indication of rank.

D: Command chariot, Tiglath-Pileser III, late 8th century BC

The command chariot of the Assyrian king was marked out by its ornate and spectacular decoration. From afar, it could be easily recognised by the unique tall parasol that shielded the monarch from the heat of the sun; this was either attached to the cab itself or carried by a eunuch. Tiglath-Pileser's chariot shows marked developments over that in Plate A. The cab has been enlarged and can accommodate up to four persons, and the wheels are substantially larger to distribute the extra weight. Development can also be seen in the appearance of the cavalryman, in this case belonging to the 'qurubti sha pitalli' or royal cavalry bodyguard. This king was the first to introduce armoured cavalrymen equipped with a bronze lamellar corselet and armed with the lance. Horses are now ridden using the proper seat.

E: Assyrian and auxiliary infantry, late 8th century BC

In this plate illustrating the appearance of infantry during the reigns of Tiglath-Pileser III and Sargon II it is possible to discern evolutionary trends in uniforms and equipment. E1 is a 'zuk shepe' or infantry guard; he wears no body armour and is very lightly equipped for an élite unit. Of note is his shield, being a cone of leather edged and embossed with bronze. E2 is a slinger, and demonstrates the form of personal body armour that was to become standard issue in later reigns. Both these native Assyrian troops lack footwear. Contact and experience with Urartian troops and the incorporation of Neo-Hittite vassal

units who employed footwear led to the Assyrians introducing such themselves; by the time of Sennacherib they are standard issue for the 'kisir sharutti'. **E3** illustrates a Neo-Hittite infantryman. The introduction of crests on Assyrian helmets dates from this period; they are heavily influenced by Urartian and Phrygian styles. A distinctive feature of Neo-Hittite troops was the 'irtu' or bronze disc protecting the chest; this feature was also appropriated by the Assyrians, and can still be seen worn by troops in the reign of Ashurbanipal.

F: Sargonid cavalry, Urartu, 714 BC

Depicted here are Assyrian horse lancers of the reign of Sargon II, on campaign against Urartu in 714 BC. These are the soldiers Sargon employed straight off the march in the battle that defeated the Urartian army to the east of Lake Urmia. Those illustrated clearly attest to the much greater proficiency of the Assyrian cavalry arm by this date. The armament is heavier, with both troopers equipped with compound bow, quiver, and long stabbing lance. The cavalry are

This bas-relief shows Assyrian cavalry conducting a campaign against Arab nomads in the desert. Fighting in such gruelling conditions illustrates once again the tactical flexibility of the Assyrian army. (British Museum)

now equipped with footwear in the form of socks with lace-up boots. Sargon speaks of attacking the mountain city of Musasir with just one thousand cavalry of this type, on his return from Urartu.

G: Sennacherib at Lachish, 701 BC

In this reconstruction of one section of the wall reliefs depicting the siege of Lachish, Sennacherib, king of Assyria, receives a report on the progress of the siege. While he is fanned by eunuchs behind his throne the report is delivered by a figure variously identified as the crown prince Ashur-nadim-shum or the senior 'turtan', Sin-ah-user. As this is done the king raises his right hand, clenching a number of arrows, in a traditional gesture of victory. A number of high-ranking officers of the royal bodyguard are to the rear of the spokesman, identified as such by the crooks on their

helmets and the maces that all carry. Of particular interest is the similarity between their uniform and that of the slingers and archers shown elsewhere in the wall relief. Nearly all are equipped with lamellar bronze body armour and tall iron conical helmet. The uniformity of such equipment points to the increasing role of the 'ekal masharti' in the provision of such items to the army. It is clear that the visual appearance of the Assyrian army during the reign of Sennacherib was very different from that even of his father, Sargon II.

H: Assyrian troops in Babylonia, early 7th century BC

In his repeated attempts to suppress the Chaldean rebels Sennacherib sent his armies into the lagoons of southern Babylonia to flush the rebels out. A 'qurubti sha pitalli' (H1) drives his horse through the water

The apotheosis of the Assyrian battle chariot is the heavy four-horse, four-man vehicle of the 7th century BC. With archer and driver covered by two shield bearers, this large and cumbersome chariot was virtually unstoppable at the charge. This relief is the inspiration for our colour Plate L. (British Museum)

while giving chase to a Chaldean rebel. Infantry support is provided by two lightly equipped Neo-Hittite auxiliaries. The infantry dress noted in the previous plate is worn in an almost identical fashion by the 'sha pitalli'. His main weapon is his long lance which he used in a downward, overarm thrusting motion; and he can employ his compound bow when needs demand, this being carried in the quiver when not employed. The infantry are unarmoured although the spearman (H2) wears a small bronze 'irtu' on his chest. The helmet bearing a chequered crest is bronze, not iron. The archer levy (H3) is totally without protection, but carries a very large capacity quiver. Although such a soldier would not normally engage in hand-to-hand fighting he carries a short stabbing sword as a defensive weapon.

I: Royal guardsman, 7th century BC

This plate illustrates the appearance of the infantry guardsmen of the reigns of Sennacherib, Esarhaddon and Ashurbanipal; while they are almost identical, there are a number of superficial differences. **I1** dates from the time of Sennacherib, and is dressed in the standard conical iron helmet with hinged cheek-pieces, lamellar armour corslet and fringed kilt. The shield is a large leather cone with a large bronze boss and decorated edging. By the time of his grandson the most significant difference in the appearance of the guardsman (**I2**) lay in the issue of these huge shields. Offering virtually whole body cover, their protective qualities must have been more than offset by the difficulty in moving such a cumbersome device in battle. The helmet has now acquired integral cheekpieces. **I3** illustrates the appearance of one of the guard officers in court dress. The richness of their apparel reflects their ceremonial role.

J: Cavalry and infantry, 655 BC

In the final stages of the battle of the Ulai River in 655 BC the Assyrian army was pushing the Elamites back into the river. Wall reliefs of the battle depict a huge mêlée with cavalry and infantry operating together. The central figure is a horse lancer, and illustrates the final appearance of cavalry before Assyria's demise. The horse is now almost fully covered by fabric armour, while in essence the trooper is little different to that in Plate H. He is supported by various Assyrian

Throughout the period of the Neo-Assyrian Empire many helmet types were worn; those worn by auxiliary and vassal troops allow tentative identification of origins. (A) Assyrian iron conical helmet, late 8th C. (B) Anatolian helmet, late 8th C. (C) Assyrian iron conical helmet with hinged earpieces, early 7th C. (D) Neo-Hittite helmet, late 8th C. (E) Assyrian iron conical helmet with integral earpieces, reign of Ashurbanipal, 7th C: the rank of 'rab-kisri' or 'shaknu' is indicated by the bronze bands.

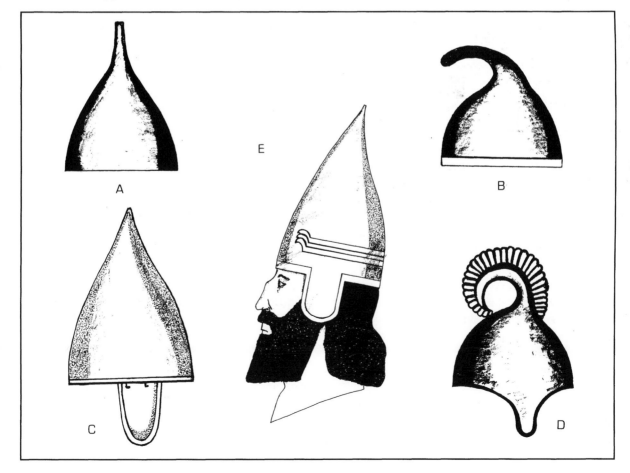

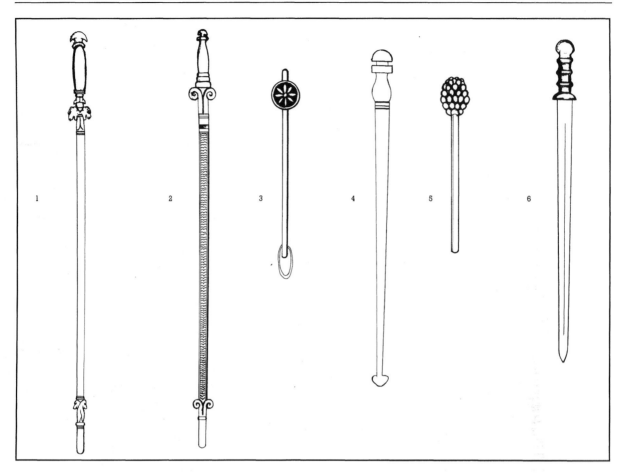

The bas-reliefs covering the Neo-Assyrian period include a multiplicity of sword types and maces. Those of the 9th C were more ornate than those of the 7th C, by which time mass production in the 'ekal masharti' arsenals led to simpler and more functional designs. (1)

Sword carried by Ashurnasirpal II, 9th C. (2) *Sword carried by cavalry trooper, 9th C.* (3) *Officer's mace, 9th C.* (4) *Infantry sword, reign of Tiglath-Pileser III, 8th C.* (5) *Officer's mace, king's bodyguard, reign of Sennacherib, early 7th C.* (6) *Infantry sword, 7th C.*

K: Heavy chariot, reign of Ashurbanipal, late 7th century BC

In its final form the Assyrian chariot is a very different machine from that shown in Plate A. It now carries four men and is pulled by four horses, and is a very heavy vehicle indeed. Such weight precluded its use in hilly country. However, at full tilt on a flat plain, as at Halule in 691 BC, it was an unstoppable and battle-winning asset. The horses are now fully covered with fabric armour, giving protection when in the charge and closing on the enemy line. The success of this design outlived the Assyrian empire, being adopted by the Babylonians under Nebuchadnezzar II.

infantry types, some of whom are wearing the standard body armour and are little different, again save for the much bigger shield reflecting a concern to avoid unnecessary losses. The lighter-equipped auxiliaries are also present. This was the final appearance of Assyrian cavalry and infantry.

INDEX

(References to illustrations are shown in **bold**. Plates are shown with caption locators in brackets.)